Introduction to Remote Sensing

Introduction
to Remote Sensing

Arthur Cracknell and Ladson Hayes

University of Dundee

Taylor & Francis
London · New York · Philadelphia

UK   Taylor & Francis Ltd, 4 John St, London WC1N 2ET

USA   Taylor & Francis Inc., 1900 Frost Road, Suite 101, Bristol, PA 19007

Copyright ©A. P. Cracknell and L. W. B. Hayes 1991

British Library Cataloguing in Publication Data

Cracknell, A. P. (Arthur Philip), *1940–*
  1. Remote Sensing
  I. Title II. Hayes, L. W. B.
  621.36′78

ISBN 0-85066-409-8
ISBN 0-85066-335-0 Pbk

Library of Congress Cataloging-in-Publication Data is available

Printed in Great Britain by Burgess Science Press, Basingstoke

Contents

Preface

We started to write this book some time ago, when there were significantly fewer elementary textbooks on remote sensing on the market than there are now. This does not exempt us from the possibility that the prospective reader or purchaser is entitled to ask why yet another elementary textbook on remote sensing is now being published. The answer is that (apart from our hope that it might be better than all the others!) we have devoted some attention to one or two things which are important and which are not usually included in elementary textbooks on this subject.

The main one of these things is the topic of atmospheric effects. We have included a whole chapter on a reasonably detailed discussion of the question of atmospheric corrections to remotely sensed data. Our reasons for this are several, but the two most important are the following. First, it is a subject in which we ourselves in the remote sensing group in Dundee University are particularly interested and on which we have done a considerable amount of research work ourselves. Secondly, and far more importantly, it is because we believe that these corrections, which are almost entirely neglected in most elementary textbooks on remote sensing, are intrinsically important in a variety of uses of remotely sensed data. This is especially so when one is trying to obtain quantitative information and particularly when one is concerned to study water areas rather than land areas; it should be remembered that water covers more than half of the surface of the Earth.

The other substantial thing that we have done is to include three appendices. The first is a brief general bibliography giving details of a number of works on the subject which it was not convenient to cite among the references in the text. For those involved in teaching remote sensing we would recommend the two books by Carter (1986) and Hyatt (1988) as extremely valuable sources of information about resource materials of all kinds for teaching purposes in remote sensing. The second appendix is a guide to sources of remotely sensed data; this is included because, in our experience, one of the questions people very frequently ask is "where can we get data from?" The third appendix is a list of abbreviations and acronyms. The remote sensing literature is full of abbreviations and acronyms and, although our list is almost certainly not complete, we do hope that it will help people with their reading of the literature.

We are grateful to the various people who have supplied us with material for illustrations and to the holders of the copyrights of material that we have used; the sources are acknowledged *in situ*. We are also grateful to Miss Mary Benstead for redrawing large numbers of the line diagrams and to the former and present staff of Taylor & Francis Ltd., especially David Grist and Robin Mellors, with whom we have been working on this project.

A. P. Cracknell

L. W. B. Hayes

1 An introduction to remote sensing

1.1 Introduction

Remote sensing may be taken to mean the observation of, or gathering information about, a target by a device separated from it by some distance. The expression "remote sensing" was coined by geographers at the US Office of Naval Research in the 1960s at about the time that the use of "spy" satellites was beginning to move out of the military sphere and into the civilian sphere. Remote sensing is often regarded as being synonymous with the use of artificial satellites and in this regard one may call to mind glossy calendars and coffee-table books of LANDSAT images of various parts of the Earth (e.g. Sheffield, 1981, 1983; Bullard and Dixon-Gough, 1985) or the satellite images that are frequently shown on television weather forecasts. Satellites do play an important role in remote sensing but it should be appreciated that remote sensing activity not only precedes the expression but dates from long before the launch of the first artificial satellite. There are a number of ways of gathering remotely-sensed data which do not involve satellites and which, indeed, have been in use for very much longer than satellites. For example, virtually all of astronomy can be regarded as being built upon the basis of remote sensing data. However, this book is concerned with terrestrial remote sensing. The idea of taking photographs of the surface of the Earth from a platform elevated above the surface of the Earth was put into practice by balloonists in the nineteenth century; the earliest known photograph from a balloon was taken of the village of Petit Bicêtre near Paris in 1859. There was substantial development of aerial photographic techniques with military reconnaissance aircraft in World War I and, much more seriously, in World War II. Later, this was accompanied by the invention and development of radar and thermal-infrared systems.

Some of the simpler instruments, principally cameras, which are used in remote sensing work also date from long before the days of artificial satellites. Any photography could be regarded as an example of remote sensing. The principle of the pinhole camera and the camera obscura has

1

been known for centuries and the photographic process for recording the image permanently on a plate, film or paper was developed in the earlier part of the nineteenth century. If remote sensing is regarded as the acquisition of information about an object without physical contact with it, almost any use of photography in a scientific or technical context may be thought of as an example of the use of remote sensing. For some decades a great deal of survey work has been done by the interpretation of aerial photography obtained from low-level flights using light aircraft; sophisticated photogrammetric techniques have come to be applied in this type of work. It is important to realize, however, that in addition to conventional photography (photography using cameras with film sensitive to light in the visible wavelength range) there are other important instruments and techniques used in remote sensing work. For instance, infrared photography can be used instead of the conventional visible wavelength range. Colour-infrared photography, which was originally developed as a military reconnaissance tool, was found to be extremely valuable in scientific studies of vegetation. Alternatively, multi-spectral scanners may be used in place of cameras. These scanners can be built to operate in the microwave range as well as in the visible, near-infrared and thermal-infrared ranges of the electromagnetic spectrum. One can also use active techniques, which are based on the principle of radar where the instrument itself generates the radiation that is used. However, the instruments used may differ very substantially from a commercially-available marine or air navigational radar.

There are other means of seeking or transmitting information apart from using electromagnetic radiation as the carrier of the information in remote sensing activities. One alternative is to use ultrasonic waves. These waves do not travel far in the atmosphere but they do travel large distances under water with only very slight attenuation; this makes them particularly valuable for use in bathymetric work in rivers and at sea, for hunting for submerged wrecks, for the inspection of underwater installations and pipelines, for the detection of fish and submarines, as well as for underwater communications purposes (see, Cracknell, 1980). Figure 1-1 shows an image of old flooded limestone mine workings obtained with underwater ultrasonic equipment.

Remote sensing is not only concerned with the generation and interpretation of data in the form of images. For instance, data on the pressure, temperature and humidity at different heights in the atmosphere are routinely gathered by the meteorological services around the world using rockets and balloons carrying expendable instrument packages which are released from the ground at regular intervals. A great deal of scientific information about the upper layers of the atmosphere is also gathered by radio sounding methods operated both from stations on the ground and from instruments flown on satellites. Close to the ground, acoustic sounding methods are often used.

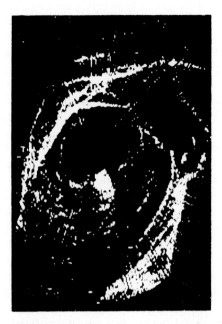

Figure 1-1 Sonar image of part of a flooded abandoned limestone mine in the West Midlands of England (Cook, 1985)

Notwithstanding the wide coverage actually implied in the term remote sensing, we shall confine ourselves for the purpose of this book to studying the gathering of information about the surface of the Earth and events on the surface of the Earth, that is we shall confine ourselves to Earth observation. This is not meant to imply that the gathering of data about other planets in the solar system or the use of ultrasound for subsurface remote sensing and communications purposes are unimportant. In dealing with the observation of the Earth's surface using remote sensing techniques, this book will be considering a part of science which not only includes many purely scientific problems, but also has important applications in the everyday lives of mankind. The observation of the Earth's surface and events thereon involves using a wide variety of instruments and platforms for the detection of radiation at a variety of different wavelengths. The radiation itself may be either radiation originating from the Sun, radiation emitted at the surface of the Earth, or radiation generated by the remote sensing instruments themselves and reflected back from the Earth's surface. A quite detailed treatise or reference book on the subject is the *Manual of Remote Sensing* (Colwell, 1983) and many details which it would not be proper to include in the present book will be found in that treatise. There are a number of general textbooks on the principles of Earth observation and its various applications and some of these are listed in Appendix I.

The original initiative and the main present driving force behind the space programme lies with the military.   The possibilities of aerial photography certainly began to be appreciated during World War I while in World War II aerial photographs obtained by reconnaissance pilots, often at very considerable risk, were of enormous importance. The use of infrared photographic film allowed camouflaged materials to be distinguished from the air.   There is little doubt that without the military impetus the whole programme of satellite-based remote sensing after World War II would be very much less developed than it is now. This book will not be concerned with the military aspects of the subject. But as far as technical details are concerned, it would seem to be a reasonably safe assumption that for any instrument or facility which is available in the civilian satellite programme there will be a corresponding instrument or facility with similar or better performance in the military programme if there is any potential or actual military need for it. As has already been indicated, the term remote sensing was coined in the early 1960s at the time that the rocket and space technology that was developed for military purposes after World War II was beginning to be transferred to the civilian domain. The history of remote sensing may be conveniently divided into two periods: the period prior to the space age, (up to 1960), and the period thereafter.   The distinctions between these two periods are summarized in Table 1-1, which is due to Colwell (1983).

Remote sensing is far from being a new technique. There was, in fact, a very considerable amount of remote sensing work done prior to 1960, although the actual term remote sensing had not been coined yet.   The activities of the balloonists in the nineteenth century and the activities of the military in World Wars I and II have already been mentioned. Following World War II enormous advances were made on the military front. Spy-planes were developed which were capable of revealing, for example, the installation of Soviet rocket bases in Cuba in 1962. Military satellites were also launched; some were used to provide valuable meteorological data for defence purposes and others were able to locate military installations and follow the movements of armies. In peacetime between World Wars I and II, substantial advances were made in the use of aerial photography for civilian applications in areas such as agriculture, cartography, forestry and geology. Subsequently archaeologists began to appreciate its potential as well. Remote sensing, in its earlier stages at least, was simply a new area in photointerpretation.   The advent of artificial satellites gave this a new dimension.

The first photographs of the Earth taken from space were released in the early 1960s. Man had previously only been able to study small portions of the surface of the Earth at one time and had painstakingly built up maps from a large number of local observations. The Earth was suddenly seen as an entity and its larger surface features were rendered visible in a way that captivated people's imagination. In 1972 the U.S.A. launched its first Earth Resources Technology Satellite ERTS-1, which was later renamed

Table 1.1   Comparison of the two major periods in the history of remote sensing

Prior to the Space Age (1860-1960)	Since 1960
A. Only one kind and date of photography	A. Many kinds and dates of remote sensing data
B. Heavy reliance on the human analysis of unenhanced images	B. Heavy reliance on the machine analysis and enhancement of images
C. Extensive use of photo interpretation keys	C. Minimal use of photo interpretation keys
D. Relatively good military/civil relations with respect to remote sensing	D. Relatively poor military/civil relations with respect to remote sensing
E. Few problems with uninformed opportunists	E. Many problems with uninformed opportunists
F. Minimal applicability of the "multi" concept	F. Extensive applicability of the "multi" concept
G. Equipment simple and inexpensive; readily operated and maintained by resource-oriented workers	G. Equipment complex and expensive; not readily operated and maintained by resource-oriented workers
H. Little concern about the renewability of resources, environmental protection, global resource information systems, and associated problems related to "signature extension", "complexity of an area's structure", and/or the threat imposed by "economic weaponry"	H. Much concern about the renewability of resources, environmental protection, global resource information systems, and associated problems related to "signature extension", "complexity of an area's structure", and/or the threat imposed by "economic weaponry"
I. Heavy resistance to "technology acceptance" by potential users of remote sensing-derived information.	I. Continuing heavy resistance to "technology acceptance" by potential users of remote sensing-derived information.

(adapted from Colwell, 1983)

LANDSAT-1. It was then imagined that remote sensing would solve almost every remaining problem in environmental science. Initially there was enormous confidence in remote sensing and a considerable degree of overselling of the new systems. To some extent, this was followed by a period of disillusionment when it became obvious that, while valuable information could be obtained, there were substantial difficulties to be overcome and considerable challenges to be met. A more realistic approach is now perceived and it is realized that remote sensing from satellites provides a tool to be used in conjunction with traditional sources of information, with aerial photography and ground observation, to improve the knowledge and understanding of a whole variety of environmental scientific, engineering and human problems.

Before proceeding any further, it is worthwhile commenting on some points that will be discussed in later sections. First, it is convenient to divide remotely-sensed material according to the wavelength of the electromagnetic radiation used, i.e. optical, near-infrared, thermal-infrared, microwave, and radio wavelengths. Secondly, it is convenient to distinguish between passive and active sensing techniques. In a passive system the remote sensing instrument simply receives whatever radiation happens to arrive and selects the radiation of the particular wavelength range that it requires. In an active system the remote sensing instrument itself generates radiation, transmits that radiation towards a target, receives the reflected radiation from the target and extracts information from the return signal. Thirdly, one or two points need to be made regarding remote sensing satellites. Manned satellite programmes are mentioned because these have often captured the popular imagination. The U.S.A. and the U.S.S.R. have for many years conducted manned satellite programmes which included cameras in their payloads. Although manned missions may be more spectacular than unmanned missions, they are necessarily of rather short duration and the amount of useful information obtained from them is relatively small compared with the amount of useful information obtained from unmanned satellites. Among the unmanned satellites it is important to distinguish between polar or near-polar orbiting satellites and geostationary satellites. Suppose that a satellite of mass m travels in a circular orbit of radius r around the Earth, of mass M, then it will experience a gravitational force of GMm/r^2 (G = gravitational constant), which is responsible for causing the acceleration r of the satellite in its orbit, where ω is the angular velocity. Thus, using Newton's second law of motion

$$G\,\frac{Mm}{r^2} \;=\; mr\omega^2 \tag{1-1}$$

or

$$\omega^2 \;=\; \frac{GM}{r^3} \tag{1-2}$$

the period of revolution, T, of the satellite is then given by

$$T = \frac{2\pi}{\omega} = 2\pi\,\sqrt{\frac{r^3}{GM}} \tag{1-3}$$

Since π, G and M are constants, the period of revolution of the satellite depends only on radius of the orbit, provided the satellite is high enough above the surface of the Earth for the air resistance to be negligible. It is very common to put a remote sensing satellite into a near-polar orbit, at about 800 – 900 km above the surface of the Earth; at that height it has a period of about 90 – 100 minutes. For a larger radius of the orbit the period will be longer. For the Moon, which has a period of about 28 days, the radius of the orbit is about 384,400 km. Somewhere in between these two radii there must be one value of the radius for which the period is exactly 24 hours or 1 day. This radius, which is approximately 42,250 km, corresponds to a height of about 35,900 km above the surface of the Earth. If one chooses an orbit of this radius and in the equatorial plane, rather than a polar orbit, and if the sense of the movement of the satellite in this orbit is the same as the rotation of the Earth then the satellite will remain vertically over the same point on the surface of the Earth (on the equator). This constitutes what is commonly known as a "geostationary" orbit.

1.2 *Aircraft* versus *satellites*

Remote sensing of the Earth from aircraft and from satellites is already established in a number of areas of environmental science. Further applications are constantly being developed as a result of improvements both in the technology itself and in people's general awareness of the potential of remote sensing techniques. Table 1-2 lists a number of areas where remote sensing is particularly useful. In the applications given, aircraft or satellite data are used as appropriate to the purpose. There are

Table 1.2   Uses of remote sensing

Archaeology and anthropology
Cartography
Geology
  surveys
  mineral resources

Land use
  urban land use
  agricultural land use
  soil survey
  health of crops
  soil moisture and evapotranspiration
  yield predictions
  rangelands and wildlife
  forestry — inventory
  forestry, deforestation, acid rain, disease

Civil Engineering
  site studies
  water resources
  transport facilities

Water resources
  surface water, supply, pollution
  underground water
  snow and ice mapping

Coastal studies
  erosion, accretion, bathymetry
  sewage, thermal and chemical pollution monitoring

Oceanography
  surface temperature
  geoid
  bottom topography
  winds, waves and currents
  circulation
  mapping of sea ice
  oil pollution monitoring

Meteorology
  weather systems tracking
  weather forecasting
  sounding for atmospheric profiles
  cloud classification

Climatology
  atmospheric minority constituents
  surface albedo
  desertification

Natural disasters
  floods, earthquakes,
  volcanoes, forest fires,
  subsurface coal fires,
  landslides

Planetary Studies

various considerations that have to be taken into account when deciding between using aircraft or satellite data. The fact that an aircraft flies so much lower than a satellite means that one can see more detail on the ground than one can see from a satellite. However, although a satellite sees less detail, it may be more suitable for many purposes. A satellite has the advantages of regularity of coverage and a scale of coverage (in terms of area on the ground) that could never be achieved from an aircraft. The frequency of coverage of a given site by satellite-flown instruments may, however, be too low for some applications. For a small area a light aircraft can be used to obtain a large number of images more frequently.

Figure 1-2 illustrates some of the major differences between satellites and aircraft in remote sensing work. There are a number of factors to be considered in deciding whether to use aircraft or satellite data; these include

- The extent of the area to be covered;

- The speed of development of the phenomenon to be observed;

- The detailed performance of the instrument available for flying in the aircraft or satellite; and

- Availability and cost of the data.

The last point in this list, which concerns the cost to the user, may seem a little surprising. Clearly it is much more expensive to build a satellite platform and sensor system, to launch it, to control it in its orbit and to recover the data than it would be to buy and operate a light aircraft and a good camera

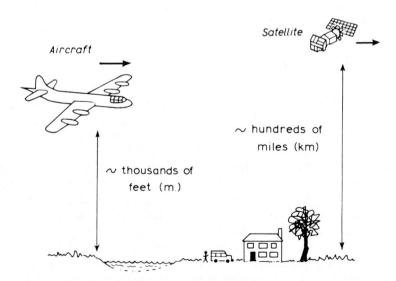

Figure 1-2 Causes of differences in scale of aircraft and satellite observations

or scanner. Until recently, however, the cost of a remote sensing satellite programme has been borne mostly by the taxpayers of one country or another and the costs charged to the user have been only related to the direct costs involved in the reception and distribution of the data. Recently however, principally as a result of changes in U.S. Government policy, there have been steep rises in the prices of many satellite data products.

The influence of the extent of the area to be studied on the choice of aircraft or satellite as a source of remote sensing data is closely related to the question of spatial resolution. Loosely speaking, we can think of the spatial resolution as being the size of the smallest object that can be seen in a remote sensing image. The angular limit of resolution of an instrument used for remote sensing work is, in nearly every case, determined by the design and construction of the instrument. Satellites are flown several hundred kilometres above the surface of the Earth whereas aircraft, and particularly light survey aircraft, may fly very low indeed, possibly only a few hundred metres above the surface of the Earth. The fact that the aircraft is able to fly so low means that, with a given instrument, far more detail of the ground can be seen from the aircraft than could be seen by using the same instrument on a satellite. However, as will be seen later, there are many purposes for which the lower resolution that is available from satellite observations is perfectly adequate and, when compared to an aircraft, a satellite can have several advantages. For instance, launched into orbit a satellite simply continues in that orbit without consuming fuel for propulsion, since the air resistance is negligible at the altitudes concerned. Occasional adjustments to the orbit may be made by remote command from the ground; these adjustments consume only a very small amount of fuel. The electrical energy needed to drive the instruments and transmitters on board satellites is derived from large solar panels.

1.3 Weather satellites

A satellite has a scale of coverage and a regularity of coverage that one could never reasonably expect to obtain from an aircraft. The exact details of the coverage obtained will depend on the satellite in question. As an example, a single satellite of the polar orbiting TIROS-N series (Television InfraRed Observation Satellite series) carries a sensor, the AVHRR (Advanced Very High Resolution Radiometer) which produces the pictures seen on many television weather programmes and which gives complete coverage of the entire surface of the Earth daily. A geostationary weather satellite gives images more frequently, as often as every half hour, but only sees a fixed portion (30-40%) of the surface of the Earth; global coverage of the surface of the Earth (apart from the polar regions) is obtained at half-hourly intervals from a chain of geostationary satellites arranged at intervals around the equator, see Figure 1-3. Satellites have completely transformed

Figure 1-3 An image of the Earth from GOES-E showing extent of geostationary satellite coverage

the study of meteorology by providing synoptic pictures of weather systems such as could never be obtained before, although in pre-satellite days some use was made of photographs from high-flying aircraft.

1.4 Observations of the Earth's surface

A satellite may remain in operation for several years unless it experiences some accidental failure or its equipment is deliberately turned off by mission control from the ground. Thus a satellite has the important advantage over an aircraft that it gathers information in all weather conditions, even those in which one might not choose to fly in a light survey aircraft. For studies of the Earth's surface it must, of course, be remembered that the surface of the Earth may be obscured by cloud. Indeed, for studies of the Earth's atmosphere it is the clouds that are often of particular interest. By flying an aircraft completely below the clouds one may be able to collect useful information about the Earth's surface although, since it is not usual for aircraft remote sensing missions to be flown in less than optimal conditions, one would try to avoid having to take aerial photographs on cloudy days. Much useful data can still be gathered by a satellite on the very large number of days on which there is some, but not complete, cloud cover.

The remotely-sensed signals detected by the sensors on a satellite or aircraft but originating from the ground are influenced by the intervening atmosphere. The magnitude of the influence depends on the distance between the surface of the Earth and the platform carrying the sensor, and on the atmospheric conditions prevailing at the time. It also depends very much on the principles of operation of the sensor, especially on the wavelength of the radiation that is used. Since the influence of the atmosphere is variable, it may be necessary to make corrections to the data in order to accommodate this variability. The approach adopted to the question of atmospheric corrections to remotely-sensed data will be determined by the nature of the environmental problem to which the data are applied, as well as by the properties of the sensor used and by the processing applied to the data. In work that has been done to date in land-based applications of remote sensing it has been relatively rare to pay much attention to atmospheric effects. In meteorological applications it is the atmosphere that is being observed anyway and quantitative determinations of, and corrections to, the radiance are relatively unimportant. Atmospheric effects have been of greatest concern to users of remote sensing data where water bodies, lakes/lochs, rivers, and the oceans, have been studied with regard to the determination of physical or biological parameters of the water.

1.5 Communications and data collection systems

1.5.1 Communications systems

Although this book is primarily concerned with remote sensing satellite platforms which carry instruments for gathering information about the surface of the Earth, mention should be made of the many satellites which are launched for use in the field of telecommunications. Many of these satellites belong to purely commercial telecommunications network operations systems. The user of these telecommunications facilities is, however, generally unaware that a satellite is being used; for example, the user simply dials an international telephone number and need never even know whether the call goes via a satellite or not. Some remote sensing satellites have no involvement in communications systems apart from the transmission back to ground of the data that they themselves generate, while others have a subsidiary role in providing a communications facility.

The establishment of a system of geostationary satellites as an alternative to using submarine cables for international communication was foreseen as early as 1945. The first communications satellite, Telstar, was launched by the U.S.A. in 1962. Telstar enabled television pictures to be relayed across the Atlantic for the short time that the satellite was in view of the ground receiving stations on both sides of the Atlantic. The Syncom series, which were truly geostationary satellites, followed in 1963. The idea involved is basically a larger version of the microwave links now commonplace on land. Two stations on the surface communicate via a geostationary satellite. The path involved is about a thousand times longer than a direct link between two stations would be on the surface. As a consequence, the antennae used are much larger, the transmitters are much more powerful, and the receivers are much more sensitive than for direct communication over shorter distances on the surface of the Earth. Further details of purely commercial telecommunications using satellites are given elsewhere (see, e.g., Cott, 1980).

For any remote sensing satellite system it is necessary to provide some means for transferring the information that has been gathered by the sensors on the satellite back to Earth. In the case of a manned spacecraft the data recorded on magnetic tape or on film can be brought back by the astronauts in the spacecraft when they return to Earth. However, the majority of scientific remote sensing data gathered from space is gathered using unmanned spacecraft. The data from an unmanned spacecraft must be transmitted back to Earth by radio transmission from the satellite to a suitably-equipped ground station. The transmitted radio signals are only able to be received from the satellite when it is above the horizon of the ground station. In the case of polar-orbiting satellites global coverage could be achieved by having tape recorders on board the satellite and transmitting the tape recorded data back to Earth when the satellite is within range of a ground station. However, in practice it is usually only possible to provide tape recording facilities adequate for recording a small fraction of the data that could, in principle, be gathered during each orbit of the satellite. Alternatively, global coverage could be made possible by the construction of a network of receiving stations suitably distributed over the surface of the Earth. This was originally intended in the case of the LANDSAT series of satellites, see Figure 1-4, but a more recent approach to securing global coverage takes the form of a relay system where a series of geostationary satellites link signals from an orbiting remote sensing satellite with a receiving station at all times.

1.5.2 Data collection systems

Although the major part of the data transmitted on the communications link from a remote sensing satellite back to Earth will consist of the data that the instruments on the satellite have gathered, some of these satellites also fulfil a communications role. For example, the geostationary satellite METEOSAT (see Section 3.2) serves as a communications satellite to transmit processed METEOSAT data from the European Space Operations Centre (ESOC) at Darmstadt in the Federal Republic of Germany to users of the data; it is also used to retransmit data from another geostationary satellite, GOES-E, to users who may be out of the direct line of sight of

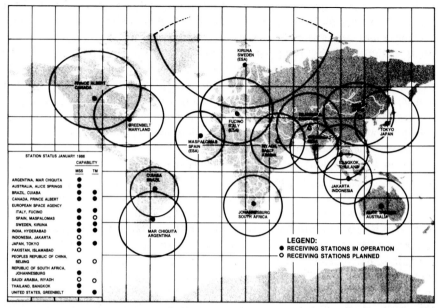

(a)

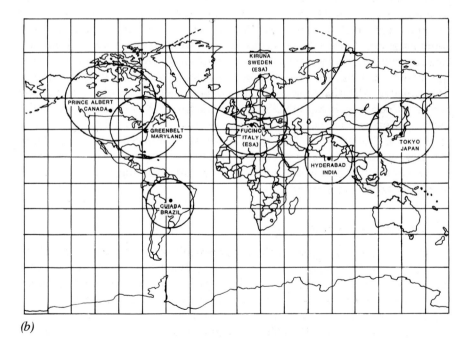

(b)

Figure 1-4 (a) LANDSAT-4, -5 MSS coverage and (b) LANDSAT-4, -5 TM coverage (station status January 1987) (EOSAT Corporation)

GOES-E itself. Another aspect of the use of remote sensing satellites for communications purposes that is of particular relevance to environmental scientists and engineers is that some satellites carry data collection systems. Such systems enable them to collect data from instruments situated in difficult or inaccessible locations on the land or sea surface. Such instruments may be at sea on a moored or drifting buoy or on a weather station in a hostile or otherwise inaccessible environment, in the Arctic, or in deserts and so on.

There are several options for recording and retrieving data from an unmanned data gathering station such as a buoy or an isolated weather station or hydrological station; these include using

1. Cassette taperecorders with occasional visits to change the tapes;

2. A direct radio link to a receiving station conveniently situated on the ground; or

3. A radio link via a satellite.

The first option may be satisfactory if the amount of data received is relatively small, but if there is a large quantity of data and one is only able to visit the site occasionally this may not be very suitable. The second option may be satisfactory over short distances but this becomes progressively more difficult over longer distances. The third option has some attractions and is worth a little further consideration here. There are two satellite-based data collection systems which are of importance. One involves the use of a geostationary satellite such as METEOSAT, while the other is the ARGOS data collection system which involves the polar-orbiting satellites of the TIROS-N, NOAA-6, -7, -8 ... series, see Figure 1-5.

There are several possible advantages in using a satellite and not just using a direct radio transmission from the platform housing the data-collecting instruments to the user's own radio receiving station. One of these is simply convenience. It saves on the cost of reception equipment and of operating staff for a receiving station of one's own; it also simplifies problems of frequency allocations. There may, however, be the more fundamental problem of distance. If the satellite is orbiting, it can store the messages on board and play them back later — perhaps on the other side of the Earth. The ARGOS system accordingly enables someone in Europe to receive data from buoys drifting in the Pacific Ocean or in Antarctica, for example. As well as recovering data from a drifting buoy the ARGOS system can also be used to locate the position of the buoy.

To some extent the METEOSAT and ARGOS data collection systems are complementary. A data collection system using a geostationary satellite, such as METEOSAT, has the advantage that the satellite is always overhead and therefore always available, in principle, to receive data. For

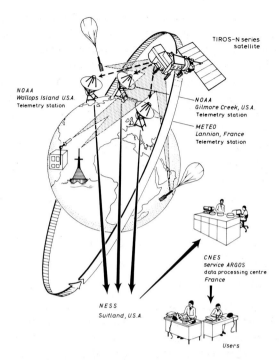

TIROS–N series
satellite

NOAA
Wallops Island U.S.A.
Telemetry station

NOAA
Gilmore Creek, U.S.A.
Telemetry station

METEO
Lannion, France
Telemetry station

CNES
service ARGOS
data processing centre
France

NESS
Suitland, U.S.A.

Users

Figure 1-5 Overview of System ARGOS data collection and platform location system (System ARGOS)

the METEOSAT system, moored buoys or stationary platforms on land can be equipped with transmitters to send records of measurements to the METEOSAT satellite; the messages are transmitted to the European Space Operations Centre at Darmstadt and then relayed to the user. Data could also be gathered from a drifting buoy using the METEOSAT system but the location of the buoy would be unknown. A data collection system cannot be used on a geostationary satellite if the data collection platform is situated in extreme polar regions, i.e. outside the circle indicating the telecommunications coverage in Figure 1-6. On the other hand a data collection system that uses a polar orbiting satellite will perform better in polar regions because the satellite will be in sight of a platform that is near one of the poles much more frequently than a platform near the equator. A polar-orbiting satellite will, however, be out of sight of the data collection platform a great deal of the time.

The platform location facility is not particularly interesting for a land-based platform because the location of the platform is known, though occasionally when transmitters have been stolen and not switched off the location facility has been a useful feature! At sea, however, information regarding the location of the data collection platform is very valuable as it allows data to be gathered from drifting buoys and provides the position

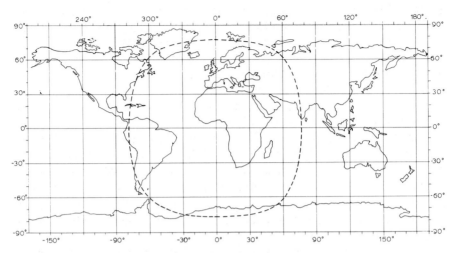

Figure 1-6 METEOSAT reception area (European Space Agency)

from which the data was obtained. The locational information is also valuable for moored buoys because it provides a constant check that the buoy has not broken loose from its mooring. If the buoy does break loose then the location facility is able to provide valuable information to a vessel sent to recover it.

The location of a platform is determined by making use of the Doppler effect on the frequency of the carrier wave of the transmission from the platform; this transmitting frequency, f_0, is fixed (within the stability of the transmitter) and (nominally) the same for all platforms. The apparent frequency of the signal received by the data collection system on the satellite will be given by

$$f' = \left\{ \frac{c - v \cos \theta}{c} \right\} f_0 \qquad (1\text{-}4)$$

where c is the velocity of light, v is the velocity of the satellite and θ is the angle between the line of sight and the velocity of the satellite. If c and f_0 are known and the orbital parameters of the satellite are known, so that v is known, then f' is measured by the receiving system on the satellite; $\cos \theta$ can then be calculated. The position of the satellite is also known from the orbital parameters so that a field of possible positions of the platform is obtained. This field takes the form of a cone with the satellite at its apex and the velocity vector of the satellite along the axis of symmetry, see Figure 1-7. A, B and C denote successive positions of the satellite when transmissions are received from the given platform. D, E and F are the corresponding positions at which messages are received from this platform in the following orbit, which will be about 90 minutes later. Since the attitude of the satellite is known, the intersection of several of the cones for one orbit (each corresponding to a separate measurement) with the altitude sphere yields the solution for the location of the platform. Actually this yields two solutions, points 1 and 1′, which are symmetrically placed relative to the ground track of the satellite. One of these points is the required solution, the other is its "image". The ambiguity cannot be resolved with data from a single orbit alone, but can be resolved with data received from two successive orbits and a knowledge of the order of magnitude of the drift velocity of the platform. In Figure 1-7 point 1′ could thus be eliminated. In practice, because of the considerable redundancy one does not require to know f_0 accurately; it is enough that the transmitter frequency f_0 be stable over the period of observation. The processing of all the measurements made at A, B, C, D, E and F then yields the platform position, its average speed over the interval between the two orbits and the frequency of the oscillator.

The ARGOS platform location and data collection system is the result of a cooperative programme between the French Centre Nationale d'Etudes Spatiales (CNES) and two U.S. organizations, the National Aeronautics and Space Administration (NASA) and the National Oceanic and Atmospheric Administration (NOAA). The ARGOS system's main mission is to provide an operational environmental data collection service for the entire duration of the NOAA TIROS-N programme, i.e. from 1979 to 1994 at least. After several years of operational service the efficiency and reliability of the ARGOS system has been demonstrated very successfully.

The ARGOS system can be considered as comprising four segments:

   1. The set of all users' platforms (buoys, balloons, fixed or

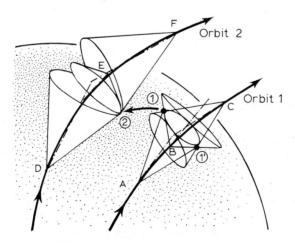

Figure 1-7 Diagram to illustrate the principle of the location of platforms with the ARGOS system

offshore stations, animals, etc.), each being equipped with a PTT (Platform Transmitter Terminal);

2. The onboard Data Collection System (DCS) flown on each satellite of the series TIROS-N, NOAA-6, -7, -8, etc.;

3. The ARGOS data-processing centre in Toulouse; and

4. The distribution system for results.

These will be considered briefly in turn.

The main characteristics of ARGOS PTTs can be summarized as follows:

- Transmission frequency : 401.640 MHz;

- Messages are of duration less than 1 second and are transmitted at regular intervals by any given PTT;

- Message capacity for sensor data: 32 to 256 bits;

- Radiated power: 3 W, for a power consumption of just 200 mW.

Since all ARGOS PTTs work on the same frequency, they are particularly easy to operate. They are also moderately priced.

At any given time, the space segment consists of two satellites equipped with the ARGOS onboard Data Collection System (DCS). These are satellites of the TIROS-N, NOAA-6, -7, -8 ... series which are in near circular polar orbits with periods of 101 minutes. Each orbit is Sun-synchronous, that is the angle between the orbital plane and the Sun direction remains constant. The orbital planes of the two satellites are inclined at $60°$ to one another. Each satellite crosses the equatorial plane at a fixed (local solar) time each day; these are 1500 hours (ascending node) and 0300 hours (descending node) for one satellite, and 1930 hours and 0730 hours for the other. These times are approximate as there is, in fact, a slight precession of the orbits from one day to the next. The PTTs are not interrogated by the DCS on the satellite — they transmit spontaneously. Messages are transmitted at regular intervals by any given platform. Time-separation of messages, to ensure that messages for different PTTs arrive randomly at the DCS on the satellite, is achieved by assigning different intervals to different platforms. Transmissions are every 40 – 60 seconds in the case of location-type platforms and every 100 – 200 seconds for data-collection-only platforms. Even if two messages do reach the satellite simultaneously, the DCS can still handle four messages at a time, provided they are separated in frequency. Frequency separation of messages will occur because the carrier frequencies of the messages from different PTTs will be slightly different as a result of the Doppler shifts of the transmissions from different platforms. Nevertheless, some messages may still be lost, but the probability of this is small and will be kept small by control of the total number of PTTs having access to the system. At any given time one of these satellites can receive messages from platforms within a circle of diameter about 5,000 km on the ground. The DCS on a satellite acquires and records a composite signal comprising a mixture of messages received from a number of PTTs within each satellite's coverage. Each time a satellite passes over one of the three telemetry stations (Wallops Island, Virginia, U.S.A.; Gilmore Creek, Alaska, U.S.A.; Lannion, France), all the ARGOS message data recorded on tape are read out and transmitted to that station. As well as being taperecorded on board the spacecraft, the ARGOS data messages are multiplexed into the direct transmissions from the satellite. Therefore a user who has a direct read-out station for AVHRR data may also receive the raw ARGOS data directly in real time from any platform that transmits while the satellite is above the horizon of the receiving station.

Once a satellite has completed telemetry data transmission for a particular pass, the received data are transmitted to the NESS (National Earth Satellite Service) centre at Suitland, Maryland (U.S.A.). Data concerning the ARGOS system are separated from the other received data and transmitted to the ARGOS Data Processing Centre (DPC) located at the Toulouse Space Centre in France.

The on-line distribution of results is available via

- The Global Telecommunications System (GTS) of the World Meteorological Organisation (WMO);

- Public switched telephone or telex networks;

- Dedicated links by telephone; and

- Data transmission networks such as Transpac in France, Euronet in Europe or Tymnet in the U.S.A.

Meteorological results are distributed as soon as processing is completed or at required times. The off-line distribution is in the form of computer compatible tape or computer printout.

The important considerations in assessing the performance of the ARGOS System are

- Data collection capability;

- Platform location accuracy; and

- Throughput time.

Regarding the data collection capability, messages appear randomly at the input to the DCS. The probability of acquisition of the messages transmitted by a given PTT during a given satellite pass over that PTT is greater than 0.99 provided that all messages transmitted during the pass are identical. The probability of a message bit error is 10^{-4} or 1 erroneous bit per 10,000. The minimum number of data-collection passes per day (for a two-satellite system) is 6 for a PTT on the Equator and up to 28 for a PTT in the polar regions. With regard to platform location accuracy, whenever location calculations are mathematically possible, various tests are performed to ensure the quality of the results. The number of location calculations per day is again a function of the latitude with a minimum of five at the Equator and a maximum of 17 in the polar regions. Location accuracy is affected by numerous factors, the most important being the medium-term stability (20 minutes) of the PTT oscillator. For high-stability oscillators the location accuracy may reach 150 m. Regarding throughput time, statistical data gathered over three years of operation revealed that, on average, 48% of all results are available to users in less than 2 hours and 30 minutes from the time of transmission by the PTT to the DCS, 61% are available in less than 3 hours and 86% are available in less than 6 hours. Thus, for example, a platform located near latitude 30° will be positioned 4 times a day and in less than 3 hours. In 1981 there were 130 users of the ARGOS system from 18 countries and a total of 578 platforms were in use; in 1982 there were 136 users and 695 platforms. In July 1986, 908 PTTs were operating: 419 on drifting buoys, 60 on moored buoys, 9 on ships, 311 fixed stations, 63 attached to animals, with 46 others devoted to miscellaneous applications. There are many applications in which the ARGOS system is employed and some of them will be mentioned briefly in Section 10.8.

2 Sensors and instruments

2.1 Introduction

Remote sensing of the surface of the Earth, whether land, sea or atmosphere, is carried out using a variety of different instruments. These instruments, in turn, use a variety of different wavelengths of electromagnetic radiation. This radiation may be in the visible, near-infrared (or reflected-infrared), thermal-infrared, microwave or radio wave part of the electromagnetic spectrum.

The nature and precision of the information that it is possible to extract from a remote sensing system depends both on the sensor that is used and on the platform that carries the sensor. For example a thermal-infrared scanner that is flown on an aircraft at an altitude of 500 m may have an instantaneous field of view (IFOV), or footprint, of about a metre square or less. If a similar instrument is flown on a satellite at a height of 800 – 900 km the IFOV is likely to be about one kilometre square. This chapter is concerned with the general principles of the main sensors that are used in Earth remote sensing. As far as the sensors which will be described in this chapter are concerned, similar sensors are, in most cases, available for use in aircraft and on satellites and no attempt will be made to draw fine distinctions between sensors developed for the two different types of platforms. Some of these instruments have been developed primarily for use on aircraft but are being used on satellites as well. Other sensors have been developed primarily for use on satellites although satellite-flown sensors are generally tested with flights on aircraft before being used on satellites. Satellite data products are popular because of their relatively low cost and because they often yield a new kind of source of information that was not available previously. For mapping to high accuracy or for the study of rapidly changing phenomena over relatively small areas, data from sensors flown on aircraft may be much more useful than satellite data.

In this chapter we shall give a brief account of some of the relevant

aspects of the physics of electromagnetic radiation (see Section 2.2)
Electromagnetic radiation is the means by which information is carried
from the surface of the Earth to a remote sensing satellite. Sensors
operating in the visible and infrared regions of the electromagnetic
spectrum will be considered in Sections 2.3 and 2.4 and sensors operating
in the microwave region of the electromagnetic spectrum will be considered
in Section 2.5. The instruments that will be discussed in sections 2.3 – 2.5
are those commonly used in aircraft or on satellites. It should be
appreciated that there are other systems being developed, operating with
microwaves and radio waves, that can be used for gathering Earth remote
sensing data using installations situated on the ground rather than in an
aircraft or on a satellite; since the physics of these systems is rather
different from that of most of the sensors flown on aircraft or satellites the
discussion of ground-based systems will be postponed until later (see
Chapter 6).

It is important to distinguish between passive and active sensors. A
passive sensor is one that simply responds to the radiation that is incident
on the instrument. In an active instrument the radiation is generated by the
instrument, transmitted downwards to the surface of the Earth, reflected
back to the sensor and the received signal is then processed to extract the
required information. As far as satellite remote sensing is concerned,
systems operating in the visible and infrared parts of the electromagnetic
spectrum are passive while microwave instruments are either passive or
active; all these instruments can be flown on aircraft as well. Active
instruments operating in the visible and infrared parts of the spectrum,
while not being flown on satellites, are flown on aircraft (see Chapter 5).
Active instruments are essentially based on some aspect of radar principles
(see Chapters 5 – 7).

Remote sensing instruments can also be divided into imaging and non-
imaging instruments. Downward-looking imaging devices produce two-
dimensional pictures of a part of the surface of the Earth or of clouds in the
atmosphere. Variations in the image field may denote variations in the
colour, the temperature, or the roughness of the area viewed. The spatial
resolution may range from a few metres, as with the synthetic aperture radar
which was flown on SEASAT, to tens of kilometres, as with the passive
scanning multi-channel microwave radiometer (SMMR) flown on SEASAT
and NIMBUS-7. Non-imaging devices give information such as the height
of the satellite above the surface of the Earth (the altimeter) or an average
value of a parameter such as the surface roughness of the sea, the wind
speed or the wind direction averaged over an area beneath the instantaneous
position of the satellite (see Chapter 7 in particular).

From the point of view of data processing and interpretation, the data
from an imaging device may be richer and easier to interpret visually, but it
usually requires more sophisticated (digital) image-processing systems to
handle it and present the results to the user. The quantitative handling of

corrections for atmospheric effects is also likely to be more difficult for imaging than for non-imaging devices.

2.2 *Electromagnetic radiation*

The important parameters characterizing any electromagnetic radiation under study are the wavelength (or frequency), the amplitude, the direction of propagation and the polarization. While the wavelength may take any value from zero to infinity, radiation from only part of this range of wavelengths is useful for the remote sensing of the surface of the Earth. First of all there needs to be a substantial quantity of radiation of the wavelength in question. A passive system is restricted to radiation that is emitted with a reasonable intensity from the surface of the Earth or which is present in reasonable quantity in the radiation that is emitted by the Sun and then reflected from the surface of the Earth. An active instrument is restricted to wavelength ranges in which reasonable intensities of the radiation can be generated by the remote sensing instrument on the platform on which it is operating. In addition to there having to be an adequate amount of radiation, it is also necessary that the radiation is not appreciably attenuated in its passage through the atmosphere between the surface of the Earth and the satellite; in other words a suitable atmospheric "window" must be chosen, see Figure 2-1.

In addition to these considerations, there is also the requirement that it must be possible to recover the data generated by the remote sensing

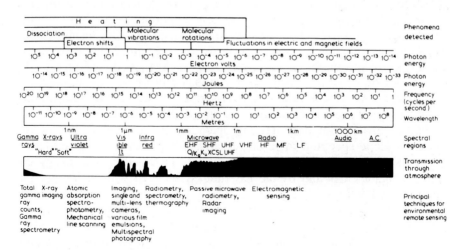

Figure 2-1 The electromagnetic spectrum. The scales give the energy of the photons corresponding to radiation of different frequencies and wavelengths (from Barrett and Curtis, 1982)

instrument. In practice this means that the amount of data generated on a satellite must be able to be accommodated both by the radio link by which the data are to be transmitted back to the Earth and by the ground receiving station used to receive the data. These various considerations restrict one to the use of the visible, infrared and microwave regions of the electromagnetic spectrum. The wavelengths involved are indicated in Figure 2-2.

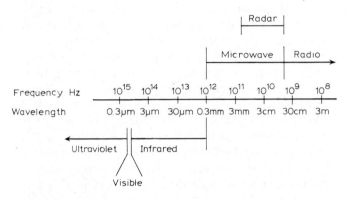

Figure 2-2 Sketch to illustrate the electromagnetic spectrum

The visible part of the spectrum of electromagnetic radiation extends from blue light with a wavelength of about 0.4 μm to red light with a wavelength of about 0.75 μm. Visible radiation will travel through a clean dry atmosphere with very little attenuation. Consequently the visible part of the electromagnetic spectrum is a very important region for satellite remote sensing work. For passive remote sensing work using visible radiation that radiation is usually derived from the Sun, being reflected at the surface of the Earth. If haze, mist, fog or dust clouds are present the visible radiation will be substantially attenuated in its passage through the atmosphere. At typical values of land-surface or sea-surface temperature the intensity of visible radiation that is emitted by the land or sea is negligibly small. Satellite systems operating in the visible part of the electromagnetic spectrum therefore usually only gather useful data during the hours of daylight. Exceptions to this are provided by aurora, by the lights of major cities and by the gas flares associated with oil production and refining activities, see Figure 2-3.

An interesting and important property of visible radiation, by contrast with infrared and microwave radiation, is that the visible radiation, especially towards the blue end of the spectrum, is capable of penetrating water to a distance of several metres. Blue light will travel 10 or 20 metres through clear ocean water before becoming significantly attenuated; red light, however, will penetrate very little. Thus with visible radiation it is

Figure 2-3 Night-time satellite image of Europe showing aurora and the lights of major cities (Aerospace Corporation)

possible to probe the physical and biological properties of the near-surface layers of water bodies, whereas with infrared and microwave radiation it will be only the surface itself that can be studied directly with the radiation.

Infrared radiation cannot be detected by the human eye, but it can be detected photographically or electronically. The infrared region of the spectrum is divided into the near-infrared, with wavelengths from about 0.75 μm to, say, about 1.5 μm and the thermal-infrared, with wavelengths from about 3 or 4 μm to about 12 or 13 μm. The near-infrared part of the spectrum is important, at least in agricultural and forestry applications of remote sensing, because most vegetation reflects strongly in the near-infrared part of the spectrum. Indeed vegetation generally reflects more strongly in the near-infrared than in the visible. Water, on the other hand, is an almost perfect absorber at near-infrared wavelengths. Apart from clouds, the atmosphere is transparent to near-infrared radiation.

At near-infrared wavelengths the intensity of the reflected radiation is considerably greater than the intensity of the emitted radiation, but at thermal-infrared wavelengths the emitted radiation becomes more important. The relative proportions of reflected and emitted radiation will vary according to the wavelength of the radiation, the emissivities of the surfaces observed and the solar illumination of the area under observation. This can be illustrated using the Planck radiation distribution function; the energy $E(\lambda)d\lambda$ in the wavelength range λ to $\lambda+d\lambda$ for black-body radiation at temperature T is given by

$$E(\lambda)d\lambda = \frac{8\pi hc}{\lambda^5[\exp(hc/k\lambda T) - 1]}d\lambda \qquad (2\text{-}1)$$

where h = Planck's constant, c = velocity of light, and k = Boltzmann's constant. The value of the quantity $E(\lambda)$, in units of $8\pi hc$ m^{-5}, is given in Table 2-1 for five different wavelengths, when $T = 300$ K, corresponding roughly to radiation emitted from the Earth. In this table $E(\lambda)(r/R)^2$ are also given for the same wavelengths, when $T = 6,000$ K, where r = radius of the Sun and R = radius of the Earth's orbit around the Sun. This gives an estimate of the order of magnitude of the solar radiation reflected at the surface of the Earth, leaving aside emissivities, atmospheric attenuation, etc.

From Table 2-1 it can be seen that at optical and very near-infrared wavelengths the emitted radiation is negligible compared with the reflected radiation. At wavelengths of about 3 or 4 μm both emitted and reflected radiation are important, while at wavelengths of 11 or 12 μm the emitted radiation is dominant and the reflected radiation is relatively unimportant. At microwave wavelengths the emitted radiation is also dominant over natural reflected microwave radiation; however as the use of man-made microwave radiation for telecommunications increases, the contamination of the signals from the surface of the land or sea becomes more serious. There is a strong infrared radiation absorption band that separates the thermal-infrared part of the spectrum into two regions or windows, one between roughly 3 μm and 5 μm and the other between roughly 9.5 μm and 13.5 μm, see Figure 2-1. Assuming the emitted radiation can be separated from the reflected radiation, satellite remote sensing data in the thermal-infrared part of the electromagnetic spectrum can be used to determine the temperature of the surface of the land or sea, provided the emissivity of the

Table 2.1 Estimates of relative intensities of reflected solar radiation and emitted radiation from the surface of the Earth

wavelength (λ)		emitted intensity	reflected intensity
blue	0·4 μm	$7·7 \times 10^{-20}$	$6·1 \times 10^{24}$
red	0·7 μm	$2·4 \times 10^{0}$	$5·1 \times 10^{24}$
infrared	3·5 μm	$1·6 \times 10^{21}$	$4·7 \times 10^{22}$
thermal-infrared	12 μm	$7·5 \times 10^{22}$	$4·5 \times 10^{20}$
microwave	3 cm	$2·6 \times 10^{10}$	$1·3 \times 10^{7}$

Note: Second column corresponds to $E(\lambda)$ in units of $8\pi hc$ m^{-5} for $T=300$ K, third column corresponds to $E(\lambda)(r/R)^2$ in the same units for $T=6000$ K

surface is known. For water the emissivity is known; it is in fact very close to unity. For the land however, the emissivity varies widely and in general its value is not known very accurately. Thus infrared remotely-sensed data can readily be used for the measurement of sea-surface temperatures, but its interpretation for land areas is more difficult. Aircraft-flown thermal-infrared scanners are widely used in surveys to study heat losses from roof surfaces of buildings as well as in the study of thermal plumes from sewers, factories and power stations, see Figure 2-4. This image highlights the discharge of warm sewage into the River Tay. It can be seen that the sewage dispersion is not particularly effective in the prevailing conditions. Since this is a thermal image the tail-off with distance from the outfall is possibly more a measure of the rate of cooling than the dispersal of the sewage.

In the study of sea-surface temperatures using the $3 - 5$ μm range, it is necessary to restrict oneself to the use of night-time data in order to avoid the considerable amount of reflected thermal-infrared radiation that is present at these wavelengths. This wavelength range is used for channel 3 of the Advanced Very High Resolution Radiometer (AVHRR) (see Section

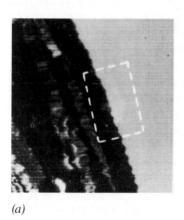

(a)

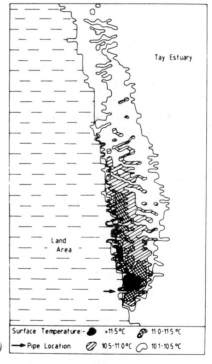

(b)

Figure 2-4 A thermal plume in the Tay Estuary, Dundee: (a) thermal infrared scanner image; (b) enlarged and thermally contoured area from within box in (a) (Wilson and Anderson, 1984)

3.4). This band of the AVHRR can accordingly be used to study surface temperatures of the Earth at night only. For the 9.5 – 13.5 µm wavelength range the reflected solar radiation is much less important and so data from this wavelength range can be used throughout the day. However, even in these two atmospheric windows the atmosphere is still not completely transparent and accurate calculations of Earth-surface temperatures or emissivities from thermal-infrared satellite data must incorporate corrections to allow for atmospheric effects. These corrections are discussed in Chapter 8. Thermal-infrared radiation does not significantly penetrate clouds, so it should be remembered that in cloudy weather it is the temperature and emissivity of the upper surface of the clouds and not of the land or sea surface of the Earth which are being studied.

In microwave remote sensing of the Earth the range of wavelengths used is from about 1 mm to several tens of centimetres. The shorter wavelength limit of this range is attributable to atmospheric absorption while the long wavelength limit may be ascribed to instrumental constraints and the reflective and emissive properties of the atmosphere and the surface of the Earth. There are a number of important differences between remote sensing in the microwave part of the spectrum and remote sensing in the visible and infrared parts of the spectrum. First, the microwaves are scarcely attenuated at all in their passage through the atmosphere, except in the presence of heavy rain. This means that microwave techniques can be used in almost all weather conditions. The effect of heavy rain on microwave transmission is actually exploited by meteorologists using special radars to study rainfall. A second difference is that the intensities of the radiation emitted or reflected by the surface of the Earth in the microwave part of the electromagnetic spectrum are very small with the result that any passive microwave remote sensing instrument must necessarily be very sensitive. This creates the requirement that the passive microwave radiometer gathers radiation from a large area (i.e., its instantaneous field of view will have to be very large indeed) in order to preserve the fidelity of the signal received. On the other hand an active microwave remote sensing instrument will have little background radiation to corrupt the signal that is transmitted from the satellite, reflected at the surface of the Earth and finally received back at the satellite. A third difference is that the wavelengths of the microwave radiation used are comparable in size to many of the irregularities of the surface of the land or the sea. Therefore the remote sensing instrument may provide data that enables one to obtain information about the roughness of the surface that is being observed. This is of particular importance in connection with oceanographic phenomena.

2.3 *Visible and near-infrared sensors*

A general classification scheme for sensors operating in the visible and

infrared regions of the spectrum is illustrated in Figure 2-5. In
photographic cameras, where an image is formed in a conventional manner
by a lens, recordings are restricted to those wavelengths for which it is
possible to manufacture lenses, i.e. in practice to wavelengths in the visible
and infrared. The camera may be an instrument in which the image is
recorded on film. Alternatively it may be like a television camera, in which
case it would usually be referred to as a Return Beam Vidicon (RBV)
camera where the image is converted into a signal which is superimposed
on a carrier wave and transmitted to a distant receiver. Return beam
vidicon cameras have been flown with some success on some of the
satellites in the LANDSAT series of satellites. In the case of the non-
photographic sensors there is either no image formed at all, or an image is
formed in a completely different physical manner from the method used in a
camera with a lens. If no lens is involved the instrument will be able to
operate at longer wavelengths in the infrared part of the spectrum.

Multi-spectral scanners are non-photographic instruments that are
widely used in remote sensing, and which are able to operate both in the
visible and infrared ranges of wavelengths. The idea of a multi-spectral
scanner involves an extension of the idea of a simple radiometer in two
ways, first by splitting the beam of received radiation into a number of
spectral ranges or "bands" and secondly by adding the important feature of
scanning. The image is not formed all at once as it is in a camera but is
built up by scanning. In most cases this scanning is achieved by using a
rotating mirror, although in some cases it is simply the whole satellite that
is spinning and in other cases a "push-broom" technique using a one-
dimensional CCD (charge-coupled detector) array is employed. A multi-
spectral scanner (MSS) consists of a telescope and various other optical and
electronic components. At a given instant the telescope receives radiation
from a given area, the instantaneous field of view (IFOV), on the surface of

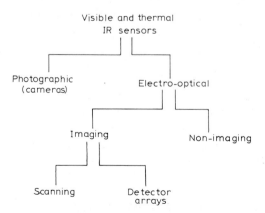

*Figure 2-5 Classification scheme for sensors covering the visible and thermal-infrared
range of the electromagnetic spectrum*

the Earth in the line of sight of the telescope. The radiation is reflected by the mirror and separated into different spectral bands, or ranges of wavelength. The intensity of the radiation in each band is then measured by a detector. The output value from the detector then gives the intensity for one point (picture element, or pixel) in the image. For a polar orbiting satellite the scanning is achieved by having the axis of rotation of the mirror along the direction of motion of the satellite so that the scan lines are at right angles to the direction of motion of the satellite, see Figure 2-6. At any instant the instrument views a given area beneath it and concentrates the radiation from that instantaneous field of view onto the detecting system; successive pixels in the scan line are generated by data from successive positions of the mirror as it rotates and receives radiation from successive IFOVs. For a polar orbiting satellite, the advance to the next scan line is achieved by the motion of the satellite. For a geostationary satellite the line-scanning is achieved by having the satellite spinning about an axis parallel to the axis of rotation of the Earth; the advance to the next scan line is achieved by adjusting the look direction of the optics, that is by tilting the mirror.

A multi-spectral scanner will produce several coregistered images, one corresponding to each of the spectral bands into which the radiation is separated by the detecting system. The number of spectral bands in a multi-spectral scanner flown on an environmental satellite is commonly four or

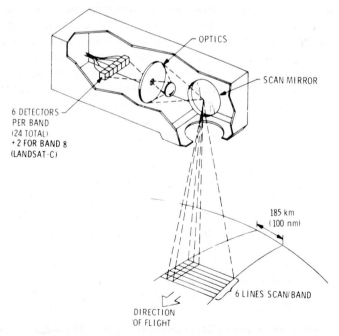

Figure 2-6 LANDSAT MSS scanning system (NASA)

five, and occasionally more. Examples are given in Chapter 3. In recent instruments a push-broom system with a one-dimensional array of CCDs (charge coupled detectors) is used in place of a scanning mirror. There is no mechanical scanning involved; a whole scan line is imaged optically onto the CCD array and the scanning along the line is achieved from the succession of signals from the responses of the detectors in the array. At a later time the instrument has been moved forwards and the next scan line is imaged on the CCD array and the responses obtained electronically, in other words the advance from one scan line to the next is achieved by the motion of the satellite. Another modern development is the extension of the number of spectral bands or channels from a single-figure number (such as 4 or 5 or 7) to several tens or even several hundreds of bands or channels. Such an instrument is described as an imaging spectrometer and the object of using more spectral bands or channels is to achieve greater discrimination between different targets on the surface of the Earth.

A great deal of information can be extracted from a monochrome image obtained from one band of a multi-spectral scanner. The image can be handled as a photographic product and subjected to the conventional techniques of photointerpretation. The image can also be handled on a (digital) interactive image-processing system and various image-enhancement operations such as contrast enhancement, edge enhancement, density slicing, etc. applied to the image. These techniques are discussed in Chapter 9. However, more information can usually be extracted by using the data from several bands and thereby exploiting the differences in the reflectivity, as a function of wavelength, of different objects on the ground. The data from several bands can be combined visually, for example, by using three bands and putting the pictures from these bands onto the three guns of a colour television monitor or onto the primary-colour emulsions of a colour film. The colours that appear in the image that is produced in this way will not necessarily bear any simple relationship to the true colours of the original objects on the ground when they are viewed in white light from the Sun. Examples of such false colour composites, see Figure 2-7, abound in many coffee table books of satellite-derived remote sensing images.

Colour is widely used in remote sensing work. In many instances the use of colour enables additional information to be conveyed visually that could not be conveyed in a black and white monochrome image, although it is not uncommon for colour to be added for purely cosmetic purposes. Combining data from several different bands of a multi-spectral scanner to produce a false colour composite image for visual interpretation and analysis suffers from the restriction that the digital values of three bands only can be used as input data for a given pixel in the image. This means that only three bands can be handled simultaneously; if more bands are used, then combinations or ratios of bands must be taken before the data are used to produce an image and in that case the information available is not being exploited to the full. Full use of the information available in all the

Figure 2-7 A false colour composite of south west Europe and north west Africa based on NOAA AVHRR data (processed by DFVLR for ESA/ESRIN)

bands can be made if the data are analysed and interpreted within a computer. The numerical methods that are used for handling multi-spectral data will be considered in some detail in Chapter 9. Different surfaces generally have different reflectivities in different parts of the spectrum. Accordingly, an attempt may be made to identify surfaces from their observed reflectivities. In doing this one needs to consider not just the fraction of the total intensity of the incident sunlight that is reflected by the surface, but also the distribution of the reflectivity as a function of wavelength. This reflectivity spectrum can be regarded as characteristic of the nature of the surface and it is sometimes described as a spectral "signature" by which the nature of the surface may be identified. However, the data recovered from the multi-spectral scanner do not provide the reflectivity as a continuous function of wavelength; one only obtains a discrete set of numbers corresponding to the integrals of the continuous reflectivity function integrated over the wavelength ranges of the various bands of the instrument, see Figure 2-8.

Thus data from a multi-spectral scanner clearly provide less scope for discrimination among different surfaces than continuous spectra would provide, but it is the best that is available. It has, in the past, not been possible to gather remotely-sensed data to produce anything like a continuous spectrum for each pixel; however the new imaging spectrometers that are being developed will, effectively, enable this to be done.

2.4 Thermal-infrared sensors

During the 1960s a versatile device known as an electro-optical linescanner was developed which provides a means of recording

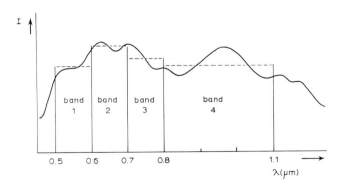

Figure 2-8 Sketch to illustrate the relation between a continuous reflectivity distribution and the band-integrated values (broken line histogram)

electromagnetic energy in the thermal region. Figure 2-9 shows a typical thermal-infrared scanning system. Radiation from the surface under investigation strikes the scan mirror and is reflected to the surface of the focusing mirrors and then to a photoelectric detector. The voltage output of the detector is amplified and activates the output of a light source. The light varies in intensity with the voltage and is recorded on film. The detectors generally measure radiation in the 3.5 – 5.5 µm and 8.0 – 14.0 µm atmospheric windows. When the instrument is operating, the scan mirror rotates about an axis parallel to the flight path (see Figure 2-9).

Infrared thermography is a passive rather than an active process. That is to say it depends on the radiation originating from the object under observation and does not require that the object is illuminated by the sensor itself. All objects with a temperature above absolute zero contain atoms in various states of random thermal motion and in continuous collision with each other. These motions and collisions give rise to the emission of

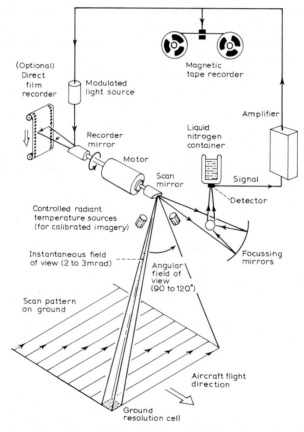

Figure 2-9 Schematic view of an infrared scanning system

electromagnetic radiation over a broad range of wavelengths. Thermography is based on the fact that the temperature of an object affects the quantity of the continuum radiation produced and determines the wavelength at which the radiation is a maximum (λ_{max}). This relationship is expressed as Wien's displacement law

$$\lambda_{max}T = \text{constant} \tag{2-2}$$

where T is the temperature of the object, and λ is the wavelength expressed in micrometres. It is not true, however, that all bodies radiate the same quantity of radiation at the same temperature. The amount bodies radiate is dependent on a property of the body called the emissivity, ϵ, the ideal black body having an emissivity of unity and all other bodies having emissivities less than unity. Wien's law describes the broadband emission properties of an object. As indicated in Section 2.2, Planck's radiation law gives the energy distribution within the radiation continuum produced by a black body.

Using the Planck relationship (Equation 2-1) it is possible to draw the shape of the energy distribution from a black body at a temperature of 293 K (20°C), the typical temperature of an object viewed by aerial thermography. The 5 to 20 μm range is also commonly referred to as the thermal-infrared region as it is in this region that objects normally encountered by human beings radiate their heat. From Figure 2-10 it can also be seen that the energy maximum occurs at 10 μm, which is fortuitous since there is an atmospheric transmission window around this wavelength. To explain what is meant by an atmospheric window, it should be realised that the atmosphere attenuates all wavelengths of electromagnetic radiation differently due to the absorption spectra of the constituent atmospheric gases. Figure 2-11 shows the atmospheric absorption for a range of wavelengths, with some indication of the gases which account for this absorption. Whereas Figure 2-11 depicts the absorption for the whole atmosphere which is of importance for satellite radiometers, Figure 2-12 shows the percentage transmittance per mile for the atmosphere close to sea level, which is a more useful measurement for aerial thermography. It can be seen then from Figure 2-11 and Figure 2-12 that there is a region of high atmospheric transmittance between 8 and 14 μm and it is this waveband which is used for thermography from airborne and satellite radiometers. This region of the spectrum is also the region in which there is maximum radiation for the range of temperatures seen in terrestrial objects (e.g. the ground temperatures, buildings and roads). The total radiation emitted from a body at a temperature T is given by the well known Stefan-Boltzmann Law

$$E = \sigma T^4 \tag{2-3}$$

Accordingly, if the total radiation emitted is measured the temperature of the body may then be determined.

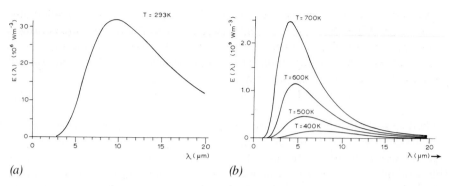

(a)   *(b)*

Figure 2-10 Planck distribution function for black body radiation at (a) 293 K, and (b) a number of other temperatures (note the change of scale between (a) and (b))

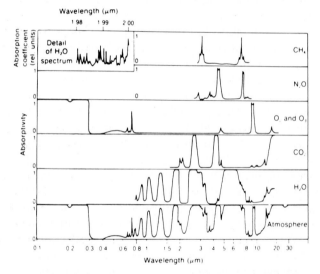

Figure 2-11 Whole atmosphere transmittance

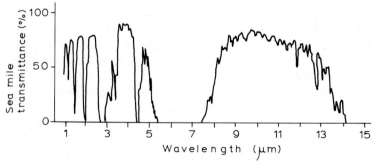

Figure 2-12 Atmospheric transmittance at sea level

Infrared surveys are flown along parallel lines at fixed line spacing and flying height and, because thermal surveys are usually flown in darkness, a sophisticated navigation system is invariably required. This may take the form of ground control beacons mounted on vehicles. Pre-dawn surveys are normally flown since the thermal conditions tend to stabilize during the night and temperature differences on the surface are enhanced. During daylight solar energy heats the Earth's surface and may accordingly contaminate the information sought. The pre-dawn period is also optimal for flying since turbulence which causes aircraft instability and consequently image distortion, is at a minimum.

The results are usually printed on conventional photographic paper to show hot surfaces as white, and cool surfaces dark, on the print. Although most interpretation is carried out using tonal and textural features, the stereoscopic effect of conventional photointerpretation may also be employed if necessary. The stereo effect is produced by side overlap of adjacent prints and not forward overlap as in conventional air photography. As a result, the stereo effect tends to be variable, and in general does not represent true or exaggerated ground-contour or surface topography.

2.5 Microwave sensors

The existence of passive microwave scanners was mentioned briefly in Section 2.2 and their advantage over optical and infrared scanners, in that they can give information about the surface of the Earth in cloudy weather, was alluded to. Passive microwave sensors also have the advantage that they are capable of gathering data at night as well as during the day since they are concerned with emitted radiation rather than reflected solar radiation. However, the spatial resolution of passive microwave sensors is very poor compared with that of visible and infrared scanners. There are two reasons for this. First, the wavelength of microwaves is much longer than visible and infrared radiation and the theoretical limit to the spatial resolution depends on the ratio of the wavelength of the radiation to the aperture of the sensing instrument. Secondly, as we have already mentioned, the intensity of microwave radiation emitted or reflected from the surface of the Earth is very low. The nature of the environmental and geophysical information that can be obtained from a microwave scanner is, of course, different from the information that can be obtained from visible and infrared scanners.

Passive microwave radiometry applied to investigations of the Earth's surface involves the detection of thermally generated microwave radiation. The characteristics of the received radiation, in terms of the variation of intensity, polarization properties, frequency and observation angle, is dependent on the nature of the surface being observed and on its emissivity. The part of the electromagnetic spectrum with which passive microwave radiometry is concerned is from ~200 GHz to ~1 GHz or, in terms of wavelengths, from ~0.15 cm to ~30 cm.

Figure 2-13 shows the principal elements of a microwave radiometer. The scanning is achieved by movement of the antenna and the motion of the platform (aircraft or satellite) in the direction of travel, see Figure 2-14. The signal is very small and one of the main problems is to reduce the noise level of the receiver itself to an acceptable level. The receiver consists basically of a high-gain antenna, a Dicke switch, a known noise source, a preamplifier, a square-law detector and an integrator. The antenna receives the very low level microwave signal. The Dicke switch is employed to switch the input to the receiver between the antenna port and a reference load which can be either a heated or cooled source, or deep space. The signal is then amplified by a low-noise preamplifier such as a tunnel diode or a parametric amplifier. After detection, the signal is integrated to give a suitable signal-to-noise value. The signal can then be stored on a taperecorder on board the platform or, in the case of a satellite, it may then be transmitted by a telemetry system to a receiving station on Earth.

The spatial resolution of a passive microwave radiometer depends on the beamwidth of the receiving antenna, the aperture of the antenna and the wavelength of the radiation

$$A_G = \frac{\lambda^2 R^2 \sec^2 \theta}{A_A} \qquad (2\text{-}4)$$

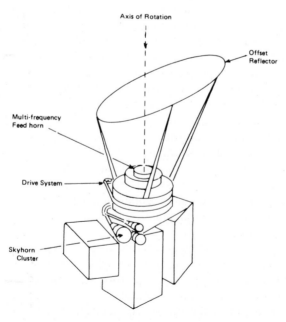

Axis of Rotation

Offset
Reflector

Multi-frequency
Feed horn

Drive System

Skyhorn
Cluster

Figure 2-13 Scanning Multi-channel (or Multi-frequency) Microwave Radiometer

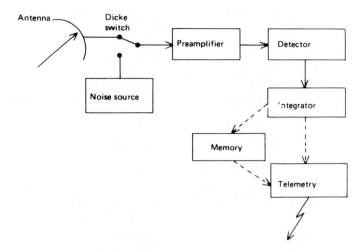

Figure 2-14 Block diagram of a microwave radiometer (Thomas, 1981)

where A_G is the area viewed (resolved normally), λ is the wavelength, R is the range, A_A is the area of the receiving aperture, and θ is the scan angle.

The spatial resolution will decrease by three or four orders of magnitude for a given size of antenna from the infrared to the microwave region of the electromagnetic spectrum. For example, the thermal-infrared channels of the AVHRR flown on the TIROS-N series of satellites have an instantaneous field of view of a little more than $1\,km^2$. For the shortest wavelength (frequency 37 GHz) of the SMMR (Scanning Multi-channel Microwave Radiometer) flown on the NIMBUS-7 satellite the instantaneous field of view is about 18 km × 27 km while for the longest wavelength (frequency 6.6 GHz) on that instrument it is about 95 km × 148 km. An antenna of a totally unrealistic size would be required to obtain an IFOV of the order of $1\,km^2$ for microwave radiation.

Passive scanning microwave radiometers flown on satellites can be used to obtain frequent measurements of sea-surface temperatures on a global scale and are thus very suitable for meteorological and climatological studies, although they are of no use in studying small-scale water-surface temperature features such as fronts in coastal regions. On the other hand the spatial resolution of a satellite-flown thermal-infrared scanner is very appropriate for the study of small-scale phenomena. It would give far too much detail for global weather forecasting purposes and would need to be degraded before it could be used for that purpose. Figure 2-15 shows sea-surface and ice-surface temperatures derived from the scanning multi-channel microwave radiometer flown on the NIMBUS-7 satellite.

There is also the problem of signal/noise ratio. The signal is the radiated or reflected brightness of the target, i.e. its microwave temperature. The noise corresponds to the temperature of the passive receiver. To improve

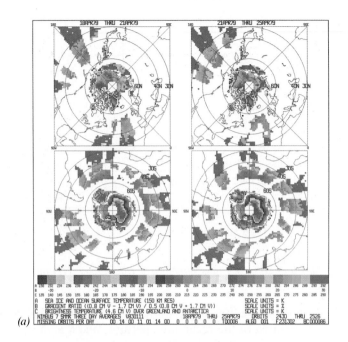

(a)

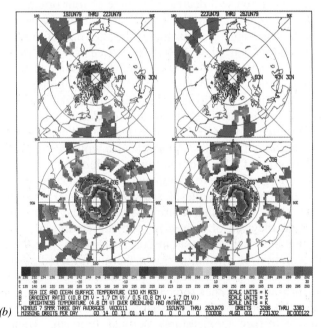

(b)

Figure 2-15 Sea ice and ocean surface temperatures derived from NIMBUS-7 Scanning Multi-channel Microwave Radiometer (SMMR) three-day average data for north and south polar regions (a) April 1979, and (b) June 1979 (NASA Goddard Space Flight Center)

the signal/noise ratio for weak targets, the receiver temperature must be proportionately lower. The signal/noise ratio, S/N, is given by

$$\frac{S}{N} = F\left(\frac{T_S^4\lambda^2}{R^2 T_R^4}\right)$$

(2-5)

where T_S is the brightness temperature of the target, T_R is the temperature of the receiver and R is the range.

The received signal in a passive radiometer is also a function of the range, the intensity of the radiation received being inversely proportional to R^2. This has a considerable effect when passive instruments are flown on satellites rather than aircraft. In practice, another important factor is the presence of microwave communications transmissions at the surface of the Earth; these are responsible for substantial contamination of the Earth-leaving microwave radiance and therefore lead to significant error in satellite-derived sea-surface temperatures.

An active microwave system can improve the poor spatial resolution associated with a passive microwave system. With an active system it is possible to measure parameters of the radiation other than just intensity. One can measure

- The time taken for the emitted pulse of radiation to travel from the satellite to the ground and back to the satellite;

- The Doppler shift in the frequency of the radiation as a result of relative motion of the satellite and the ground; and

- The polarization of the radiation (although the polarization can also be measured by passive instruments).

The important types of active microwave instruments that are flown on satellites include the altimeter, the scatterometer and the synthetic aperture radar (SAR).

A radar altimeter is an active device which uses the return time of a pulse of microwave radiation to determine the height of the satellite above the surface of the land or sea. It is the vertical distance straight down from the satellite to the surface of the Earth which is measured. Altimeters have been flown on Skylab, GEOS-3 and SEASAT and accuracies from ± 30 cm to ± 10 cm have been obtained with them. The principal use of the altimeter is for the determination of the (mean) level of the surface of the sea after the elimination of tidal effects. It is also possible, by analysing the shape of the return pulse received by the altimeter when the satellite is over the sea, to determine the significant wave height of waves on the surface of the sea and to determine the near-surface wind speed, but not the wind direction. The relationships used to determine the sea state and the wind speed are essentially empirical relationships. These empirical relationships are based originally on measurements obtained with altimeters flown on aircraft and calibrated with surface data; subsequent refinements of these relationships have been achieved using using satellite data. Accuracies of ± 2 ms^{-1} are claimed for the derived wind speeds.

The scatterometer is another active microwave instrument that can be used to study sea state. Unlike the altimeter which uses a single beam directed vertically downwards from the spacecraft, the scatterometer uses a more complicated arrangement that involves a number of radar beams which enable the direction as well as the speed of the wind to be determined. It was possible to determine the wind direction to within $\pm 20°$ with the scatterometer on the SEASAT-1 satellite. The values obtained were average values over areas of approximately 50 km^2. Further details of active microwave systems are presented in Chapter 7.

The important imaging microwave instruments are the passive scanning multi-channel (multi-spectral, or multi-frequency) microwave radiometers and the active synthetic aperture radars. It has already been noted that passive radiometry is limited by its poor spatial resolution which depends on the range, on the wavelength of the radiation used, on the aperture of the antenna and on the signal/noise ratio. The signal/noise ratio in turn is influenced by the strength of the signal produced by the target and by the temperature and sensitivity of the receiver. Ideally, a device is required that can operate in all weather conditions, that can operate both during the day and during the night and that has adequate spatial resolution for whatever purpose it is required to use the instrument in an Earth observation programme. For many remote sensing applications passive microwave radiometers cannot satisfy the third requirement. An active microwave instrument, that is some kind of radar device, meets the first two of these conditions, the conditions concerning all-weather and night-time operation. When used on an aircraft, conventional imaging radars are able to give very useful information about a variety of phenomena on the surface of the Earth. Accordingly, conventional (side-looking airborne) radars (SLARs) are frequently flown on aircraft for remote sensing work. However, when it comes to carrying an imaging radar on board a satellite, calculations of the size of antenna that would be required to achieve adequate spatial resolution show that one would need an antenna that was enormously bigger than one could possible hope to mount on board a satellite. Synthetic aperture radar (SAR) has been introduced to overcome this problem. In a synthetic aperture radar, reflected signals are received from successive positions of the antenna as the platform moves along its path. In this way an image is built up that is similar to the image one would obtain from a real antenna of several hundreds of metres or even a few kilometres in length. Whereas in the case of a radiometer or a scanner, an image is produced directly and simply from the data transmitted back to Earth from the platform, in the case of a synthetic aperture radar the reconstruction of an image from the transmitted data is much more complicated. This involves the processing of the Doppler shifts of the received radiation and will be described in Chapter

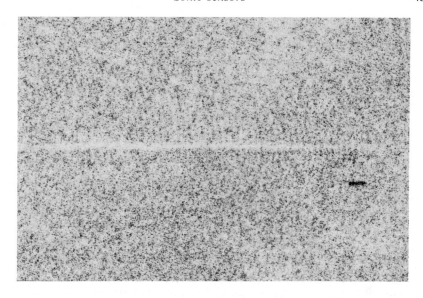

Figure 2-16 Displacement of a ship relative to its wake in a SAR image; data from SEASAT orbit 834 of 24 August 1978 processed digitally (RAE Farnborough)

7. It is important not to have too many preconceptions about the images produced from a synthetic aperture radar. A synthetic aperture radar image need not necessarily be a direct counterpart of an image produced in the optical or infrared part of the spectrum with a camera or scanner. Perhaps the most obvious difference arises in connection with moving objects in the target field. Such an object will lead to a received signal that has two Doppler shifts in it, one from the motion of the target and one from the motion of the platform carrying the synthetic aperture radar instrument. In the processing of the received signals it is not possible to distinguish between these two different contributions to the Doppler shift. Effectively, the processing will regard the Doppler shift arising from the motion of the target as an extra contribution to the range. Figure 2-16 is a synthetic aperture radar image of a moving ship in which the ship appears displaced from its wake; similarly synthetic aperture radar images have been obtained in which a moving train appears displaced sideways from the track. The principles of synthetic aperture radar are considered in Chapter 7.

2.6 Sonic sensors

2.6.1 Sound navigation and ranging (SONAR)

Sonar is similar in principle to radar but uses pulses of sound or of ultrasound instead of pulses of radio waves. Whereas radio waves propagate freely in the atmosphere but are heavily attenuated in water, the opposite is true of ultrasound. Radar cannot be used under water. Sonar is used very extensively for underwater studies, both for ranging, for detecting underwater features and for mapping seabed topography. The underwater features may include wrecks, or in a military context, submarines and mines.

2.6.2 Echo sounding

Two methods are generally available for observing seabed topography. In the first, and more familiar, method, an echo sounder is used to make discrete measurements of depth below floating vessels, and from profiles of such measurements water depth charts can be constructed. This is, essentially, the underwater analogue of the radar altimeter used to measure the height of a satellite above the surface of the Earth. The echo sounder method gives a topographic profile along a section of the sea floor directly beneath the survey ship. Even if a network of such lines is surveyed, considerable interpolation is required if the echo sounder data are to be contoured correctly and a meaningful two-dimensional picture of seabed topography constructed between traversed lines.

Echo sounders do not provide direct measurement of water depth. A pulse of sound is emitted by the sounder and the echo from the seabed is detected. It is the time interval between transmission of the pulse and detection of the echo that is measured. This pulse of sound has travelled to the seabed and back over a time interval called the two-way travel time. Thus the depth d is given by

$$d = \tfrac{1}{2}vt \tag{2-6}$$

where t = the two-way travel time, and v = velocity of sound in water. The velocity v is not a universal constant but its value will depend on factors such as the temperature and salinity of the water.

The first stage in the production of bathymetric charts from echo soundings is the transferral of depth values measured for each fix position onto the survey map, the depth values being termed 'posted' values. Depth values intermediate between fixes are usually posted at this stage, particularly topographic highs and lows as seen on the echo trace. Once a grid of lines has been surveyed in an area, the data may be contoured to produce a bathymetric chart. However, it is first necessary to apply corrections to the measured depth values to compensate for tidal effects, to adjust to a predefined datum and to compensate for variation with depth of the velocity of sound in water.

2.6.3 Side scan sonar

The side scan sonar was developed in the late 1950s from experiments using echo sounders tilted away from the vertical. Such sounders were

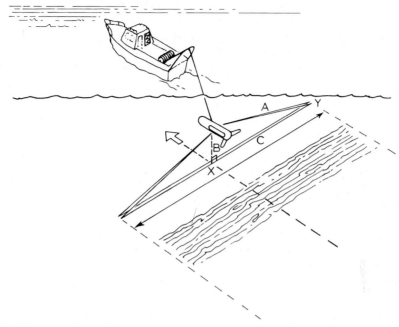

Figure 2-17 Artist's impression of a side scan sonar transducer beam: A, slant range; B, towfish height above bottom; C, horizontal range (actual distance between X and Y) (Klein Associates Inc.)

studied as a possible means of detecting shoals of fish, but results also showed the potential of the method for studying the geology of the seabed and the detection of wrecks as well as natural features of seabed topography adjacent to, but not directly beneath, a ship's course. Modern equipment utilizes specially designed transducers which emit a focused beam of sound having a narrow horizontal beam angle, usually less than 2°, and a wide vertical beam angle, usually greater than 20°; each pulse of sound is of very short duration, usually less than 1 msec. To maximize the coverage obtained per survey line sailed, systems have been designed which are dual-channel, the transducers being mounted in a towed "fish" so that separate beams are scanned to each side of the ship, (see Figure 2-17). Thus a picture can be constructed of the seabed ranging from beneath the ship to up to a few hundred metres either side of the ship's course. The range of such a system is closely linked to the resolution obtainable. Emphasis is given here to high resolution, relatively short range systems (100 m – 1 km per channel) as these are more commonly used. Typically, a high precision system would be towed some 20 m above the seabed and would survey to a range of 150 m on either side of the ship.

As with the echo sounder, the basic principle is that of detecting echoes of a transmitted pulse and presenting these on a facsimile record, termed a sonograph, in such a way that the time scan can easily be calibrated in terms of distance across the seabed. The first echo in any scan is the bottom echo, subsequent echoes being reflected from features ranging across the seabed to the outer limit of the scan. A number of points should be noted. The range scale shown on a sonograph is usually not the true range across the seabed but the slant range of the sound beam, A in Figure 2-17, and, as with the echo sounder, distances indicated on a record depend on an assumption as to the velocity of sound in water, since the distance is taken to be equal to $\frac{1}{2}vt$. If properly calibrated, the sonograph will show the correct value for B, the depth of water beneath the fish, which is presented as a depth profile on a sonograph. Echoes reflected across the scan, subsequent to the seabed echo, (from points X to Y in Figure 2-17), are subject to slant range distortion: the actual distance scanned across the seabed is

$$c = \sqrt{A^2 - B^2} \qquad\qquad (2\text{-}7)$$

Thus, corrections for slant range distortion should be applied if an object is detected by side scan and it is required to make a precise measurement of its size and position relative to a fixed position of a ship. If A and B are kept near constant for a survey, a correction can be made to relate apparent range to true range. Most of the distortion is in the zone immediately beneath the towed fish. In the range $50 - 150$ m distortion is negligible. However, if the fish was to be towed at 50 m above the seabed the distortion would be more significant over a wider part of the scan, and uncorrected range values would differ significantly from corrected values over the full range of the scan. In practical terms, if it is possible to tow the sonar fish above the seabed in the vicinity of $10 - 20$ per cent of the full-scale setting chosen for operation, then the sonograph will be free from excessive non-linear slant range distortion except in the region very close to the line of the profile. Sonar fish housings are usually designed on the assumption that these conditions will be met, and the transducers are set with the vertical beam tilted down at $10°$ from horizontal. In some respects, scale distortion is a more important type of distortion to which sonographs are subject. Distances measured along the sonograph are usually not equal to distances measured across the scan direction. In some cases it is possible to obtain near isometric presentation by varying the paper drive speed, but not all equipment has this facility, and often the quality of the record would be worsened if such a solution were attempted. If it is necessary to make mosaics of the seabed in the same way as in aerial photography, then records must either be photographed using a non-linear photographic technique, or side scan data must be recorded on magnetic tape or disk to enable processing to provide an isometric presentation at a selected mapping scale through a display or other output device. Both oscilloscope and fibre-optic recorder based systems have been developed for this.

Perhaps the most important variable to be considered in side scan sonar applications is the resolution required. For highest resolution, a high

Figure 2-18 Side scan sonar system components (Klein Associates Inc.)

frequency sound source, possibly in the range 50 – 500 kHz, and a very short pulse length, of the order of 0.1 msec, is required. Such a source gives a range resolution of 20 – 50 cm and enables detection of small-scale features of seabed morphology such as sand ripples of 10 – 20 cm amplitude. However, the maximum range of such a system is not likely to exceed 400 m. If lower resolving power is acceptable, systems based on lower frequency sources are available which can be operated over larger sweep ranges. Thus, if the object of a survey is to obtain complete coverage of an area, range limitation will be an important factor in the cost of the undertaking.

The configuration of the main components of the instrument system is very similar to that of the echo sounder, though with dual channel systems each channel constitutes a single channel sub-system consisting of transmission unit, transmitting and receiving transducers, receiving amplifier and signal processor, see Figure 2-18. Both channels then feed a common recording, display and time control unit. The sonar transmission unit is designed to produce a high voltage pulse of short duration to activate the transmitting transducer. Capacitance discharge techniques have been developed for this purpose. The pulse causes the transducer to oscillate at a high frequency at a large power level, but over only a few cycles. Transmitting and receiving transducers are constructed as line arrays of piezoelectric elements and in a dual channel system four such arrays are located in the tow fish.

The design and characteristics of both the tow cable and the fish have a large effect on the quality and flexibility of operation of the complete system. In the cable, receiver leads need to be individually screened to eliminate crosstalk between channels and the cable needs to have a high breaking strain rating, yet be reasonably flexible for deck handling — a few hundred metres of cable being required for deep tow operations. The tow fish must also swim without yawing as this would distort the sonar signals and the tow fish/cable assembly must be capable of depression to a depth of within a few tens of metres of the seabed at survey speed and be towed at this depth along an aproximately horizontal plane. These requirements have been met in different ways by different manufacturers, but all involve fairly difficult handling procedures, a situation often exacerbated by the fact that the sonar fish is an expensive item to lose through collision with the seabed.

The function of receiving amplifier and signal processor in a side scan sonar is similar to that of the equivalent unit in the echo sounder, but as it is not just those signals due to the first arrival echoes from the seabed that are of concern, a more complex signal-processing facility is required. In particular, time variable gain is necessary to ensure that signals due to echoes from the far range of a sonar scan are printed on the facsimile record with the same order of intensity as those echoes due to echoes from targets close to the sonar fish. In recent years, considerable advances have been made in signal-processing techniques in this regard. Early sonar instruments suffered from a number of defects which made consistency in record quality difficult to obtain with variation of water depth and seabed conditions, both across scan and along the profile.

The recording and display unit, with associated timing mechanism, is again similar in operation and function to echo sounders, except that in being a dual channel system in most instruments a means has to be devised for split-trace or dual-trace printing so that both channels are printed onto the same roll of recording paper. One way of meeting this requirement is to use a recorder fitted with helical electrodes wound with opposite pitch on the same drum. Sonar records are fix marked in the same way as echo sounding traces, and the position of important targets can be established by applying corrections for slant range distortion, if this distortion is significant.

3 Satellite systems

3.1 Introduction

In April 1960, only 3.5 years after the first man-made satellite orbited the Earth, the United States began its environmental satellite programme with the launch of the first satellite in its TIROS (Television and InfraRed Observation Satellite) series. This was the first in a long series of satellites launched primarily for the purpose of meteorological research. An enormous number of other satellites have been launched since then for environmental remote sensing work.

In this chapter a few of the important features of some remote sensing satellite systems are outlined. Rather than attempt to describe every system that has ever been launched, this chapter will concentrate on those which are reasonably widely used by those scientists and engineers who actually make use of remote sensing data collected by satellites. These include

- GOES-E, GOES-W and METEOSAT;

- The LANDSAT-1, LANDSAT-2, LANDSAT-3 series;

- The TIROS-N series;

- Experimental satellites such as HCMM, the NIMBUS series, especially NIMBUS-7, and SEASAT; and

- New systems such as LANDSAT-4, LANDSAT-5, SPOT, MOS and ERS-1.

Most of the satellites have been launched by the United States. The exceptions in this list are the European geostationary satellite METEOSAT, the French SPOT satellite, the Japanese MOS satellite, and the planned ESA satellite ERS-1.

The first group of satellites are geostationary satellites, while the remainder are polar-orbiting satellites. The main features of the multi-spectral scanners and other sensors on these systems are summarized and data on one or two scanners that have been flown on aircraft are included for comparison in Table 3-1. Data from the DMSP (U.S. Defense Meteorological Satellite Program) are also available and are very similar to the data obtained from the TIROS-N series of satellites.

Consideration is given to the spatial resolution, spectral resolution and frequency of coverage of the different systems. While it is also important to consider the question of atmospheric effects on radiation travelling from the surface of the Earth to a satellite, as they do also influence the design of the remote-sensing systems themselves, an extensive discussion of these is postponed until Chapter 8, as the consideration of atmospheric effects is of importance primarily at the data processing and interpretation stages.

3.2 *GOES-E, GOES-W and METEOSAT*

A geostationary satellite can be positioned over any point on the equator. The objective has been to place a geostationary satellite at about every 60° or 70° around the equator. Given the horizon that can be seen from the geostationary height this gives global coverage of the Earth with the exception of the polar region, (see Figure 3-1). The satellites at 75°W (GOES-E) and at 135°W (GOES-W) are operated by the U.S.A. METEOSAT, a European meteorological satellite, is positioned over the Greenwich meridian. METEOSAT spins at 100 rpm about an axis almost parallel to the N-S axis of the Earth (see Figure 3-2). Changes in inclination, spin rate and longitudinal position are made, when required, by using a series of thruster motors which are controlled from the ground. The main instrument on board the satellite is a scanning radiometer which has three spectral bands, (see also Table 3-1):

1. Visible (VIS) $0.4 - 1.1$ μm;

2. Thermal infrared (IR) $10.5 - 12.5$ μm;

3. Infrared water vapour absorption (WV) $5.7 - 7.1$ μm.

The third wavelength range is a little unusual; this band indicates atmospheric water vapour content.

The IFOV at the equator is a square of side 2.5 km for the visible and 5 km for the IR and WV channels. An IR or WV image consists of 2500 lines × 2500 pixels, or picture elements. A visible image consists of 2500 lines × 5000 pixels; there are two visible detectors and they are offset by one scan line so that a full visible image using both detectors would give 5000 lines × 5000 pixels. The WV channels share the sampling electronics of one visible channel and therefore every thirty minutes one obtains either

- IR + both VIS channels; or

- IR + one VIS + WV channels.

METEOSAT is also used for communications purposes, (see section 1.5).

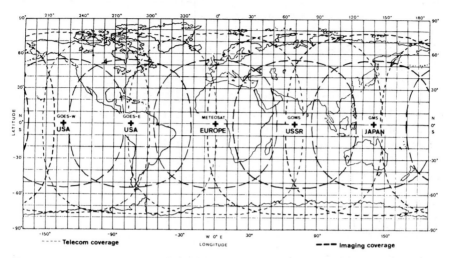

Figure 3-1 Coverage of the Earth by the international series of geostationary meteorological satellites

The GOES-E and GOES-W satellites operated by the U.S.A. are very similar to METEOSAT, but they only have two, rather than three, spectral channels and the IFOV is rather larger than for METEOSAT.

3.3 LANDSAT-1, LANDSAT-2 and LANDSAT-3

Each of these satellites was placed in a near-polar Sun-synchronous orbit at a height of about 918 km above the surface of the Earth. The satellite

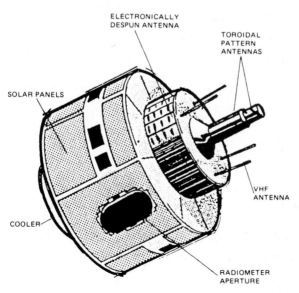

Figure 3-2 The first METEOSAT satellite

Table 3.1 Features of some commonly-used multispectral scanners

LANDSAT – 1, – 2, – 3 MSS

Channel	Wavelength (μm)	IFOV (m)
4	0·5 – 0·6	80
5	0·6 – 0·7	80
6	0·7 – 0·8	80
7	0·8 – 1·1	80

NIMBUS – 7 CZCS

Channel	Wavelength (μm)
1	0·433 – 0·453
2	0·51 – 0·53
3	0·54 – 0·56
4	0·66 – 0·68
5	0·6 – 0·8
6	10·5 – 12·5
Resolution	~825m

METEOSAT

Channel	Wavelength (μm)	IFOV (km)
1	0·4 – 1·1	~2·4
2	10·5 – 12·5	~5
3	5·7 – 7·1	~5

HCMM

Channel	Wavelength (μm)	IFOV (km)
1	0·5 – 1·1	0·5
2	10·5 – 12·5	0·5

Ocean Colour Scanner (OCS) flown on aircraft during EURASEP North Sea Experiment

Channel	Centre wavelength (nm)	Full Bandwidth at half intensity (nm)	IFOV (m)
1	431	24·2	~70 when flown at altitude of 19·8 km
2	472	26·0	
3	506	25·0	
4	548	26·3	
5	586	24·1	
6	625	25·3	
7	667	24·2	
8	707	26·0	
9	738	24·0	
10	778	26·1	

TIROS-N AVHRR

Channel	Wavelength (μm)	IFOV (km)
1	0·55 – 0·9	1·1
2	0·725 – 1·0	1·1
3	3·55 – 3·93	1·1
4*	10·5 – 11·5	1·1

SPOT

Channel	Wavelength (µm)	IFOV (m)
Multi-spectral mode	0·5 − 0·59	20
	0·61 − 0·68	20
	0·79 − 0·89	20
Panchromatic mode	0·51 − 0·73	10

LANDSAT — 4 Thematic Mapper

Channel	Wavelength (µm)	IFOV (km)
Blue	0·45 − 0·52	30
Green	0·52 − 0·6	30
Red	0·63 − 0·69	30
Near-infrared	0·76 − 0·9	30
Mid-infrared	1·55 − 1·75	30
Infrared	2·1 − 2·35	30
Thermal infrared	10·4 − 12·5	120

ERS-1 ATSR

Wavelength (µm)	IFOV (km)
3·7	1
11	1
12	1

SEASAT VHRR

Channel	Wavelength (µm)	IFOV (km)
1	0·49 − 0·94	3
2	10·5 − 12·5	5

Separated into two channels (10·3-11·3 µm and 11·5-12·5 µm) in later versions of the instrument.

travels in a direction slightly west of south and passes overhead at about 1000 hours local solar time (LST). In a single day there will be 14 southbound (daytime) passes (the northbound passes are at night), see Figure 3-3. The distance between successive paths is much greater than the swath width, see Figure 3-4, and so not all the Earth is scanned in any given day. The swath width is 185 km and, for convenience, the data from each path of the satellite is divided into frames or scenes corresponding to tracks on the ground approximately 185 km; each of these scenes contains 2286 scan lines and there are 3200 pixels on each scan line. The orbit is precessing slowly so that on the following day all the paths will be moved slightly to the west and it will only be after 18 days, i.e., on the 18th day, that the pattern repeats itself exactly. There is some overlap of orbits and in northerly latitudes this overlap becomes quite large.

Each of the first three LANDSAT satellites carried a multi-spectral scanner, MSS, and return beam vidicon cameras. The MSS on LANDSAT-3 had an additional thermal-infrared band, but it generated relatively little useful data. Otherwise these instruments all involve four bands in the visible or near-infrared part of the electromagnetic spectrum, see Table 3-1. These bands are commonly labelled 4, 5, 6 and 7 instead of 1, 2, 3 and 4, although with LANDSAT-4 the more logical numbers 1, 2, 3 and 4 have now been introduced. The spectral responses for the bands normalized to a common peak are sketched in Figure 3-5.

At these wavelengths the surface of the Earth will be obscured if there is cloud present. Thus considering the fact that there will be only one or two passes in 18 days (16 days for LANDSAT-4 and LANDSAT-5) and that there may be cloud on the day of overpass, it can be seen that the number of

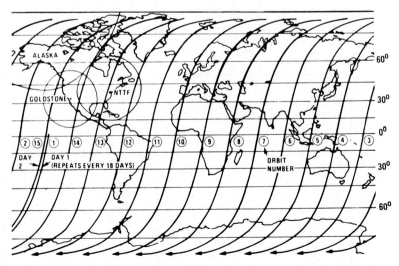

Figure 3-3 LANDSAT-1, -2 and -3 orbits in one day (NASA 1976)

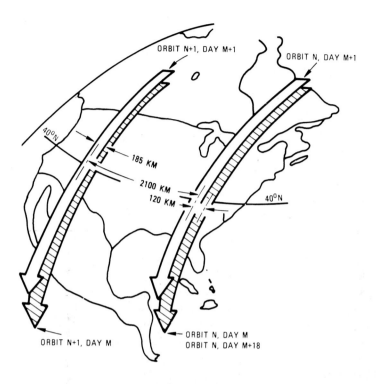

Figure 3-4 LANDSAT-1, -2 and -3 orbits over a certain area on successive days (NASA 1976)

useful LANDSAT passes per annum over a given area might be fewer than half a dozen. Nonetheless, data from the multi-spectral scanners on the LANDSAT series of satellites have been used very extensively in a large number of remote sensing programmes. As their name suggests, the LANDSAT satellites were designed primarily for remote sensing of the land, but in certain circumstances useful data are also obtained over the sea and inland water areas.

3.4 TIROS-N series

The satellites in this series, TIROS-N, NOAA-6, NOAA-7, NOAA-8, NOAA-9, NOAA-10, ... have several instruments on board; of these the most well known is the Advanced Very High Resolution Radiometer (AVHRR). This is a multi-spectral scanner, like those of the LANDSAT series, but with a very much poorer spatial resolution, namely about 1 km,

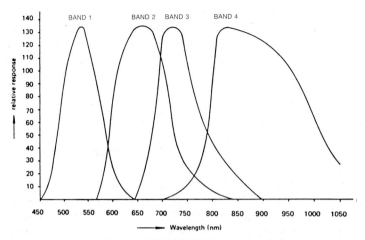

Figure 3-5 LANDSAT MSS wavelength bands

(see Table 3-1). The swath width is correspondingly wider, about 3000 km or more compared with 185 km in the case of the LANDSAT multi-spectral scanner. The swath width for the AVHRR is sufficiently wide to produce overlap between the areas scanned from successive orbits and so daily coverage of the whole of the Earth is provided. It is the AVHRR that is often used to give the satellite images used by weather forecasters on TV.

The AVHRR has four spectral bands although the actual wavelengths are different from the LANDSAT MSS wavelengths, see Table 3-1. In the later version of the instrument the thermal-infrared fourth band has been replaced by two bands at 10.3 – 11.3 μm and 11.5 – 12.5 μm. The main reason for introducing an extra infrared band is to enable the production of more accurate charts of sea-surface temperatures by using the data from several bands to eliminate atmospheric effects.

The AVHRR is enormously successful for meteorological and oceanographic purposes where the spatial resolution is perfectly adequate for the purpose. For coastal and estuarine work, however, the spatial resolution is often not sufficiently fine. The AVHRR can also, of course, be used for some purposes over land areas, provided its spatial and spectral resolution are appropriate. For instance it has been used in connection with the determination of snow cover, and hence melt-water run-off forecasting, for the Norwegian hydro-electric power board (Østrem, 1981). Considerable interest is developing in the use of the AVHRR for large area surveys of vegetation, water resources and land form (Hayes, 1985; Justice *et al.*, 1985). However, the AVHRR is not the only instrument on the TIROS-N series of satellites; there are several others which are often neglected, possibly because they are non-imaging sensors. These instruments include the TIROS Operation Vertical Sounder, TOVS, which itself is a three-sensor atmospheric sounding system, the Space Environment Monitor, SEM, which measures solar particle flux at the spacecraft, and the ARGOS data collection system (see Chapter 1). The data from the TOVS can be used to assist in the atmospheric correction of AVHRR data.

3.5 HCMM, NIMBUS-7 and SEASAT

These three satellite systems were all experimental satellites carrying very different payloads.

The HCMM carried a two-channel scanner with rather better spatial resolution in the thermal-infrared range than had been achieved with previous satellite-flown scanners, (see Table 3-1 for details). The principal objective of this programme was to obtain temperature measurements of the Earth's surface at 12-hourly intervals and hence to deduce the thermal inertia. The mission data were obtained during the period from April 1978 to September 1980 and are available from NASA Goddard Space Flight Center and from EARTHNET.

NIMBUS-7 and SEASAT were both launched in 1978. SEASAT only lasted about three months, but NIMBUS-7 continued to operate for nearly ten years. The two-channel optical and infrared scanner on SEASAT, (see Table 3-1), was of rather low spatial resolution and was of little importance in comparison with the scanners on other satellites. On NIMBUS-7 the important instruments were

- The Scanning Multi-channel Microwave Radiometer, (SMMR); and

- The Coastal Zone Colour Scanner, (CZCS).

On SEASAT the more important instruments were

- The altimeter;

- The scatterometer;

- The Synthetic Aperture Radar (SAR); and

- The Scanning Multi-channel Microwave Radiometer (SMMR).

All the SEASAT sensors and the SMMR on NIMBUS-7 are microwave sensors; the SMMR has already been described in Section 2.5 while the active microwave instruments, the altimeter, scatterometer and synthetic aperture radar, will be described in Chapter 7. The CZCS on NIMBUS-7 is an optical and infrared multi-spectral scanner which has proved to be extremely important; it is similar in many ways to the LANDSAT MSS and to the AVHRR. The instantaneous field of view (IFOV) of the CZCS is comparable with that of the AVHRR. The CZCS has six spectral channels, including some very narrow channels in the visible and a thermal-infrared channel, see Table 3-1. The CZCS spectral bands in the visible are particularly appropriate for marine and coastal work, although it might be argued that for near coastal work the IFOV is rather large. The frequency of coverage of the CZCS is more like that of the AVHRR than that of the LANDSAT MSS but NIMBUS-7 has power budget limitations and so the CZCS is only switched on for relatively short periods that fall very far short of the full 90 or so minutes of the complete orbit. Thus even if one knows, from orbit considerations, that it is possible to obtain an image of a certain area at a certain time with the CZCS there is no guarantee that the instrument will actually be switched on at that time by NASA. While the spatial resolution of the CZCS is much the same as the spatial resolution of the AVHRR, the instrumental noise in the thermal-infrared band of the CZCS is much worse than in the thermal-infrared band of the AVHRR.

3.6 New satellite systems

This section briefly considers some recently-launched and future satellite systems. These include the new satellites in the LANDSAT series (LANDSAT-4 onwards) and the first French SPOT satellite launched on 22 February 1986. The new LANDSAT satellites carry a multi-spectral scanner similar to those carried on earlier satellites in the LANDSAT series and also a much-improved multi-spectral scanner, the Thematic Mapper, see Table 3-1. This instrument has seven spectral bands and a spatial resolution with an IFOV of only 30 m in the visible bands and near and middle-infrared bands and an IFOV of only 120 m in the thermal-infrared band. The sensor on the French SPOT series of satellites has three spectral bands with an IFOV of 20 m in the multi-spectral mode or only 10 m in the panchromatic mode, (see Table 3-1), — even better spatial resolution than the Thematic Mapper. The SPOT sensor is of the push-broom type and this arrangement allows longer signal integration time which serves to reduce instrumental noise. However, unlike the case with a scanner, there is a need to calibrate the detectors across each scan line.

The ESA Remote Sensing Satellite ERS-1 is due for launch in 1991. ERS-1 is to be particularly relevant to marine applications of remote sensing. The main part of the payload will be a set of active microwave instruments which will be similar to those that were flown on SEASAT. There is, however, going to be an additional sensor. This is the Along Track Scanning Radiometer, ATSR/M, which is an infrared imaging instrument with some additional microwave channels proposed by a team from the Rutherford Appleton Laboratory at Oxford. It is designed for accurate sea-surface temperature determination. This instrument will use three spectral bands in the infrared and will scan vertically downwards and at 60° from nadir in the forward direction, so that from two positions of the satellite that are only very slightly separated in time two views are obtained of a given element of area on the surface of the sea. The object of this is to enable atmospheric effects to be eliminated to a high order of accuracy. There is a trade off between radiometric accuracy and spatial resolution. In areas of open ocean, the aim is to achieve an absolute accuracy of ±0.5 K in sea-surface temperature determinations at a spatial resolution of about 50 km in areas of open ocean, and to work with better spatial resolution (*c* 1 km IFOV) in coastal areas with a relative accuracy of ±0.5 K. The first mode of operation is aimed at climate research, medium to long-term weather forecasting and certain oceanographic applications, while the second is aimed at coastal zone oceanographic applications, fisheries, and other smaller-scale applications where the finer spatial resolution is essential and absolute accuracy of the temperature can be sacrificed. The ATSR/M will also have two microwave channels at 23.8 and 36.5 GHz to provide accurate determination of the total water vapour content of the atmospheric column beneath the satellite.

3.7 Resolution

In discussing remote sensing systems, there are three important and related quantities that need to be considered, namely

    1. Spectral resolution;

    2. Spatial resolution (or instantaneous field of view); and

    3. Frequency of coverage.

Each of these quantities is briefly considered in turn in this section with particular reference to multi-spectral scanners, see Table 3-1; Many of the ideas involved apply to other imaging systems such as radars and even to some non-imaging systems too. Spectral resolution is determined by the construction of the sensor system itself, while the instantaneous field of view and frequency of coverage are determined both by the construction of the sensor system and by the conditions under which it is flown. To some extent there has to be a trade-off between spatial resolution and frequency of coverage; good spatial resolution, i.e. small IFOV, tending to be associated with low frequency of coverage, (see Table 3-2).

3.7.1 Spectral resolution

It has already been mentioned that it is not possible to obtain measurements of reflectivity as a continuous function of wavelength for all the pixels in an image obtained with a multi-spectral scanner. All that is obtainable are integrated reflectivities over the wavelength bands used in the scanner. For most land-based applications of multi-spectral scanner data from satellites the number of visible and invisible spectral bands of the

Table 3.2   Frequency of coverage versus spatial resolution

System	IFOV	Repeat coverage
SPOT multispectral	20 m	days – variable*
panchromatic	10 m	
LANDSAT MSS	80 m	several days‡
TM	30 m	
NOAA AVHRR	~ 1 km	few hours‡
METEOSAT	~ 2·5 km	30 minutes

*pointing capability complicates the situation
‡exact value depends on various circumstances

LANDSAT MSS is adequate. For some coastal and marine applications, for example, in the determination of suspended sediment loads and chlorophyll concentrations, as many spectral channels as possible are required. In this regard the LANDSAT MSS is poor while CZCS is tolerable. For other applications, such as sandbank mapping and sea-surface temperature determinations, there is less need for a multitude of spectral channels. For sea-surface temperature determinations, only one appropriate infrared channel is required and this, by and large, has been available on existing scanners. However, additional channels may be very valuable in correcting for or eliminating atmospheric effects. The use of the split channel in the thermal-infrared region on the later versions of the AVHRR has already been mentioned in this regard. The SMMR has five spectral channels and is also able to discriminate between two polarization directions for the radiation used, thereby giving effectively ten channels. With these ten channels it is possible to eliminate atmospheric effects quite successfully. For detecting oil pollution at sea the panchromatic band of the AVHRR or the visible bands of LANDSAT MSS would seem to be adequate from the spectral point of view.

3.7.2 Spatial resolution

For land-based applications within large countries such as the U.S.A., Canada, China, the U.S.S.R., etc., the spatial resolution of the LANDSAT MSS, with its IFOV of approximately 80 m, is adequate for many purposes. For land-based applications on a finer scale, however, the spatial resolution of the LANDSAT MSS is not as good as one might like; this situation is improved substantially as data from the Thematic Mapper on the

LANDSAT series and from SPOT are now available with spatial resolutions of 30 m and 20 m (or 10 m) respectively.

In coastal and estuarine work the spatial resolution of the LANDSAT MSS is adequate for many purposes. The spatial resolution of other instruments is not; even a quite wide estuary is quickly crossed within half a dozen, or fewer, pixels for AVHRR, CZCS or HCMM.

For oceanographic work the spatial resolution of the AVHRR, CZCS or the scanners on the geostationary satellites is generally adequate. The IFOV of the AVHRR or the CZCS is of the order of 1 km^2. For the METEOSAT radiometer the spatial resolution is considerably poorer because the satellite is very much higher above the surface of the Earth; the IFOV is about 5 km square for the thermal-infrared channel of METEOSAT. At the other extreme, of course, for a thermal-infrared scanner flown in a light aircraft at a rather low altitude the IFOV may be only 1 m^2; aerial surveys using such scanners are now widely used to monitor heat losses from the roofs of large buildings. In areas of open ocean the spatial resolution of the AVHRR, CZCS or METEOSAT radiometer is perfectly adequate — it provides the oceanographer with synoptic maps of sea-surface temperatures over enormous areas; these could not be obtained on such a scale in any other way before the advent of remote sensing satellites. Such maps are also beginning to find uses in marine exploitation and management, e.g. in assisting in the location of fish, and in marine and weather forecast modelling.

3.7.3 Frequency of coverage

There is a fairly simple trade-off between spatial resolution (or instantaneous field of view, IFOV) and frequency of coverage. At a given stage in the development of the technology the constraints imposed by the sensor design, the on-board electronics and the data link to the ground will limit the total amount of data that can be obtained. Thus the smaller the IFOV, the more data there is to be handled for any given area on the ground and the less frequently data will be available for a given area, (see Table 3-2). For the LANDSAT MSS the spatial resolution is rather less than 80 m but coverage is only once or twice in 16 days. For the AVHRR or CZCS the spatial resolution is reduced by about an order of magnitude to about 1 km, but the frequency of coverage is about an order of magnitude better, of the order of twice a day or more. However, it is not just a matter of instrument specifications and orbit considerations that limit the frequency of coverage as there are also considerations associated with both the platform power requirements and the actual reception and recovery of the data. As mentioned previously, the CZCS requires too much power to be left switched on for a complete orbit, with the result that as well as needing an overpass of a particular area to obtain data there is also the requirement that the instrument be switched on. A similar problem associated with power requirements applied to the synthetic aperture radar that was flown on SEASAT.   With that system it was not possible to obtain global coverage by using on-board taperecorders because the rate of data output from the sensor was too high to enable it to be recorded on board at all. Accordingly, SEASAT synthetic aperture radar data could only be recovered for areas near to the ground receiving stations situated in the U.S.A., Canada and Great Britain that were able to receive the data directly. On the other hand, data from the low data-rate instruments on SEASAT were taperecorded on board and recovered on a global basis.

4 Data reception, archiving and distribution

The philosophy behind the gathering of remote sensing data is rather different in the case of satellite data and in the case of aircraft data. Aircraft-flown data are usually gathered in a campaign that is commissioned by, or on behalf of, a particular user and is carried out in a predetermined area. Aircraft data are usually also gathered for a particular purpose such as making maps or monitoring some given natural resource. The instruments, wavelengths and spatial resolution used will be chosen to suit the purpose for which the data are to be gathered. The owner of the remotely sensed data may or may not decide to make the data generally available.

The philosophy behind the supply and use of satellite remote sensing data, on the other hand, is rather different and the data are likely to be gathered on a speculative basis. The organisation or agency that is involved in launching a satellite, controlling the satellite in orbit and recovering the data gathered by the satellite is not necessarily the main user of the data and is unlikely to be operating the satellite system on behalf of a single user. Frequently it is the practice not to collect data only from the areas on the ground for which there is known in advance to be a customer for the data. But, rather, data are often collected over enormous areas and archived for subsequent supply when users later identify their requirements. A satellite system is usually established and operated by an agency of a single country or by an agency involving collaboration among the governments of a number of countries. Apart from the task of actually building the hardware of the satellite systems and collecting the remotely-sensed data, there is the task of archiving and disseminating the data and, indeed, of convincing the potential end-user community of the relevance and importance of the data to their particular needs.

In Chapter 2 we have described a number of different instruments and satellite systems. The problem of recovering, on the surface of the Earth, the data generated by these systems is a problem in telecommunications. The output signal from an instrument, or a number of instruments, on board a spacecraft is superimposed on a carrier wave and this carrier wave, at radio-frequency, is transmitted back to Earth.

Let us consider a particular example. On each of the TIROS-N/NOAA series of satellites there is a set of instruments including:

- AVHRR, the advanced very high resolution radiometer;

- HIRS/2, the high resolution infrared radiation sounder;

- SSU, the stratospheric sounding unit;

- MSU, the microwave sounding unit;

- SEM, the space environment monitor; and

- DCS, the ARGOS data collection and platform location system.

The AVHRR is a multi-spectral scanner that generates images of enormous areas at a spatial resolution of about 1 km (see Chapter 3). Consequently it generates data at a very high rate, namely 665,400 bps (bits per second) or 0.6654 Mbs (megabits per second). All the other instruments produce very much smaller quantities of data. The HIRS/2, SSU and MSU are known collectively as the TIROS Operational Vertical Sounder (TOVS) and are used to gather atmospheric data. The SEM is concerned with measuring solar proton, alpha particle and electron flux density, energy spectrum and the total particulate energy disposition at the altitude of the satellite. The ARGOS data collection system has already been mentioned in Section 1.5.2. All these five instruments generate very small quantities of data in comparison with the AVHRR, (see Figure 4-1). It will be seen that the data rates of these instruments range from 2,880 bps to 160 bps, which is to be compared with 665,400 bps for the AVHRR.

The TIROS-N/NOAA series of satellites are operated with three separate transmissions, the Automatic Picture Transmission (APT), the High Resolution Picture Transmission (HRPT) and the Direct Sounder Broadcast (DSB). Figure 4-1 identifies the frequencies used and attempts to indicate the data included in each transmission. The HRPT is an S-band transmission at 1698.0 or 1707.0 MHz and includes the data from all the instruments and the spacecraft housekeeping data. For the APT transmission a degraded version of the AVHRR data is produced; this consists of data from only two of the five spectral bands and the ground resolution (IFOV) is degraded from about 1 km to about 4 km. Although this means that the received picture will be of poorer quality than the full-resolution picture obtained with the HRPT, the advantage of the APT is that the transmission can be received with simpler equipment than one needs for the reception of the HRPT. The DSB transmission contains only the data from the low data-rate instruments and does not include even a degraded form of the AVHRR data.

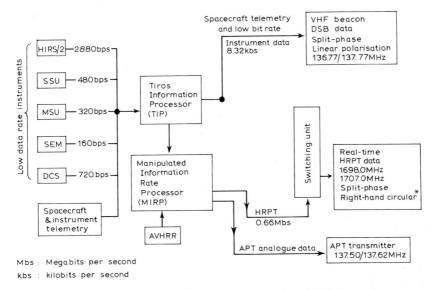

Figure 4-1 TIROS-N instrumentation (NOAA)

While the higher-frequency transmissions contain more data, there is a price to be paid in the sense that both the data-reception equipment and the data-handling equipment need to be more complicated and more expensive. Thus, to receive the S-band HRPT transmission requires a large and steerable reflector/antenna system instead of just a simple fixed antenna. In addition to having the machinery to move the antenna it is also then necessary to have quite accurate information about the orbits of the spacecraft so that the antenna assembly can be pointed in the right direction so as to be ready to receive transmissions as the satellite comes up over the horizon. Thereafter the assembly has to be moved so that it continues to point at the satellite as it passes over. The other important consequence of having a high data-rate is that more complicated and more expensive equipment will be needed to accept and store the data while the satellite is passing over.

For the TIROS-N/NOAA series of satellites the details of the transmission are published. The formats used for arranging the data in these transmissions and the calibration procedure for the instruments as well as the values of the necessary parameters are all published too (Lauritson *et al.*, 1979). Anyone is free to set up the necessary receiving equipment to recover the data and to use the data. Indeed NOAA has for a long time adopted a policy of positively encouraging the establishment of local receiving facilities for the data from this series of satellites. A description of the equipment required to receive HRPT and to extract and archive the data is given by Baylis (1981, 1983) based on the experience of the facility established at Dundee University, see Figure 4-2. It is now

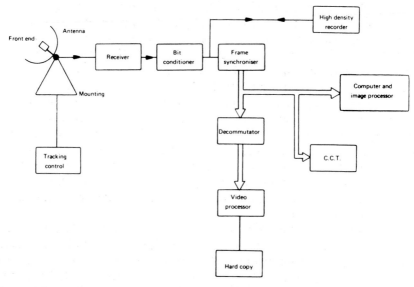

Figure 4-2 Block diagram of a receiving station for AVHRR data (Baylis, 1981)

possible to buy "off-the-shelf" systems for the reception of satellite data from commercial suppliers.

It should be appreciated that it is only possible to receive radio transmissions from a satellite while that satellite is above the horizon as seen from the position of the ground reception facility. Thus, for the TIROS-N/NOAA series of satellites the area of the surface of the Earth for which AVHRR data can be received by one typical data reception station, namely that of the French Meteorological Service at Lannion in North West France, is shown in Figure 4-3. For a geostationary satellite the corresponding area is very much larger because the satellite is very much farther away from the surface of the Earth (see Figure 1-6). Thus, although one can set up a receiving station to receive direct readout data, if one wishes to obtain data from an area beyond the horizon — or to obtain historical data — one has to adopt another approach. One may try to obtain the data from a reception facility for which the target area is within range. Alternatively, one may be able to obtain the data via the reception and distribution facilities provided by the body responsible for the operation of the satellite system in question for historical data and data from areas beyond the horizon. Consider, for example, the TIROS-N/NOAA series of satellites. These satellites carry taperecorders on board and so NOAA is able to acquire imagery from all over the world. In addition to the real-time, or direct-readout, transmissions which have just been described some of the data obtained in each orbit is taperecorded on board the satellite and played back while the satellite is within sight of one of NOAA's own ground stations at Wallops Island, Virginia, and Gilmore Creek, Alaska, in the U.S.A. It is only possible in this way to recover a small fraction of all the data obtained in an orbit. The scheduling and playback are controlled from the NOAA control room (see Needham, 1983). The data are then archived on magnetic tapes and distributed in response to requests from users. In a similar way each LANDSAT carries tape recorders which allow global coverage of data; the data are held by the EROS Data Center. Governmental space agencies which have launched remote sensing satellites, such as NASA in the U.S.A. or ESA in Europe, have also established receiving stations, both for receiving data from their own satellites and also for receiving data from other satellites as well. The data transmitted from civilian remote sensing satellites are not usually encrypted and the technical information on transmission frequencies and signal formats is usually available.

The radio signals transmitted from a remote sensing satellite can, in principle, be received not just by the owner of the spacecraft but by anyone who has the appropriate receiving equipment and the necessary technical information. In the case of the TIROS-N/NOAA series of satellites, as we have seen already, the necessary technical information on the transmission and on the formatting and calibration of the data is readily available and

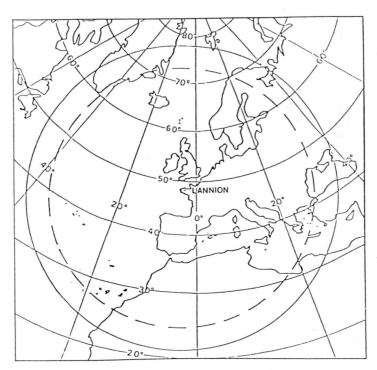

Figure 4-3 Lannion (France) NOAA polar-orbiting satellites data acquisition zone

there are no restrictions on the reception, distribution and use of the data. However, it should not be assumed that the same is true for all remote sensing satellites, or even for all civilian remote sensing satellites. For example, the situation with regard to LANDSAT is quite different from that for the TIROS-N/NOAA series. The receiving hardware needs to be more sophisticated as the data rate is higher than for the meteorological satellites; moreover, to operate a facility for the reception and distribution of LANDSAT data it is necessary to pay a large annual licence fee. LANDSAT ground receiving stations are established in various parts of the world; in the U.S.A. (Goldstone, Greenbelt, Fairbanks), Europe (Fucino, Kiruna, Mas Palomas), Canada (Prince Albert), India (Hyderabad), Brazil (Cuiaba), Argentina (Mar Chiquita), South Africa (Hartebeeshoek), Japan (Tokyo), Australia (Alice Springs), Thailand (Bangkok), China (Beijing) and Saudi Arabia (Riyadh), while others are planned in Africa (Nairobi, Kinshasa, Ouagadougou), New Zealand (Otaki) and Chile (Santiago), see Figure 1-4. European LANDSAT data are distributed by EARTHNET, a part of ESA (the European Space Agency). Originally all the European data were received at Fucino in Italy, but more recently the reception, archiving and distribution of data for northern Europe has been transferred to Kiruna in Sweden. Distribution is via the National Points of Contact in the member states of ESA. Addresses of the principal sources of remotely sensed data are given in Appendix II.

5 Lasers and active aircraft remote sensing systems

5.1 Introduction

As mentioned in Chapter 1, it is convenient to distinguish between active and passive systems in remote sensing work. The vast majority of the work that has been done at optical wavelengths involves passive techniques and all the work that is done at infrared wavelengths involves passive techniques, while at microwave wavelengths both passive and active techniques are important. The passive optical work involves both conventional photographic cameras, possibly used with selections of filters, and multi-spectral scanners, see Section 2.3. However, there is a small but important part of optical remote sensing work that involves active techniques based on the use of lasers. Light bounced off a satellite from lasers situated on the ground is used to carry out ranging measurements to enable the precise determination of the orbit of a satellite. However, the use of lasers mounted on a remote sensing platform above the surface of the Earth is, for the moment at least, restricted to aircraft. Lasers are not used on free-flying satellites because they require large collection optics and extremely high power.

In any application of active optical remote sensing techniques, i.e. of lasers, it is either of two principles that is involved. The first involves the use of the lidar principle, that is the radar principle applied in the optical region of the electromagnetic spectrum. The second involves the study of fluorescence spectra induced by a laser. These techniques have so far been applied mainly in a marine context, with lidar being used for bathymetric work in rather shallow waters and fluorosensing being used for hydrocarbon pollution monitoring. NASA has developed an Airborne Oceanographic Lidar (AOL) which was intended to be capable of operating both in a bathymetric mode and in a fluorosensing mode (Hoge and Swift, 1980, 1981; Hoge et al., 1980).

5.2 Lidar bathymetry

The charting of the foreshore and inshore shallow-water areas is one of the most difficult and time-consuming aspects of conventional hydrographic surveying from boats. This is because there is a requirement for closely-packed sounding lines, which necessitates a large amount of data collection as each sounding line represents a sampling over a very narrow swath.   In addition to the constraint of time, shallow-water surveying presents the constant danger of surveying boats running aground.

Attempts have been made to use the (passive) multi-spectral scanner (MSS) data from the LANDSAT series of satellites for bathymetric work in shallow waters (Cracknell *et al.*, 1982*a*; Bullard, 1983*a,b*) but there are a number of problems that arise (see, for example, MacPhee *et al.*, 1981). These problems are attributable to the various contributions to the intensity of the light over a water surface reaching a scanner flown on a satellite, see Figure 5-1. The use of MSS data for water-depth determination is based on mathematical modelling of the total radiance, of all wavelengths, received at the scanner and the subtraction of the unwanted components leaving only those attributable to water depth, see Figure 5-2.   By subtracting atmospheric scattering and water-surface glint, the remaining part of the received radiance is due to the "water-leaving radiance". This water-leaving radiance arises from diffuse reflection at the surface and from radiation which has emerged after travelling from the surface to the bottom and back again; the contribution of the latter component will depend on the water absorption, the bottom reflectivity and the water depth.   The feasibility of extracting a measured value of the depth depends accordingly on being able to separate these factors. The separation is made easier by the fact that the absorption and scattering of light by water is wavelength-dependent. In the most simple case the comparison of the radiance received from two wavelengths which have similar reflectivities at the bottom but different absorption properties will yield depth directly. However, a more general depth algorithm using varying wavelengths is needed to account for changes in water quality. For example, with data gathered from the estuary of the River Tay in Scotland, contours have been drawn down to about 5 m depth with a standard deviation of about 0.5 m (Cracknell *et al.*, 1982*a*). There is, however, a very high concentration of suspended sediment in the Tay; in clearer waters it may be possible to use MSS data for bathymetric work to greater depths, e.g., work has been done at depths down to about 20 m in the Red Sea (Bullard, 1983*b*). Apart from the limitation on the depth to which the technique can be used there is the further difficulty that the horizontal spatial resolution of the MSS on the LANDSAT series of satellites is rather poor for bathymetric work in shallow waters.   The problem of spatial resolution, as well as that of atmospheric correction, can be reduced by using scanners flown on aircraft instead of satellites but, even so, it seems unlikely that sufficient accuracy for charting purposes will always be obtainable. An appreciable increase in performance could be realized by combining the MSS with an active system such as a scanning laser bathymeter (see Figure 5-3) which would provide a constant calibration for the removal of effects due to variable water quality.

A method for carrying out bathymetric surveys involving conventional aerial colour photography in association with a laser system has been developed by the Canadian Hydrographic Service in cooperation with the

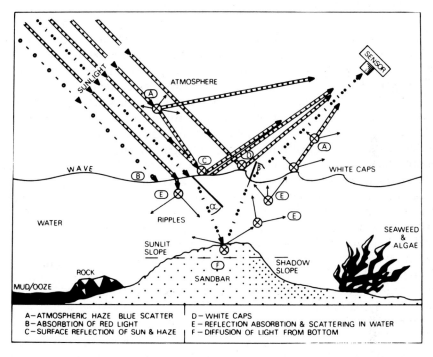

Figure 5-1 Light intensity reaching a satellite (Bullard, 1983a)

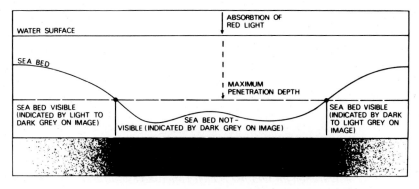

Figure 5-2 Depth of water penetration represented by a grey scale (Bullard, 1983a)

Canada Centre for Remote Sensing. This development, which began in 1970, consists of a photohydrography system and a laser profiling system. The photohydrography system uses colour photography while the laser system uses a profiling laser bathymeter. Both systems are flown simultaneously. The photography provides 100% bottom coverage over a depth range of 2 – 10 m for typical sea water and other information, such as shoreline, shoals, rock outcroppings and other hazards to navigation. The laser system utilizes a single-pulsed laser transmitter and two separate receivers, one to receive the echoes back from the surface of the water and the bottom, the other to measure aircraft height. The laser which is used transmits short high-power pulses of green light (532 nm) and infrared radiation (1064 nm) at a repetition rate of 10 Hz, and two optical/electronic receivers, one tuned to 532 nm and the other to 1064 nm, are employed to detect the reflected energy, see Figure 5-4. The green light penetrates the water rather well, while the infrared radiation hardly penetrates the water at all. Echoes from the surface and from the bottom are received by the green channel and from these the water depth is obtained by measuring the difference in echo arrival times. Aircraft height information is acquired by the infrared channel which measures the two-way transit time of each 1064 nm pulse from the aircraft to the water surface. The lidar bathymeter is used to provide calibration points along a line or lines so that depths can

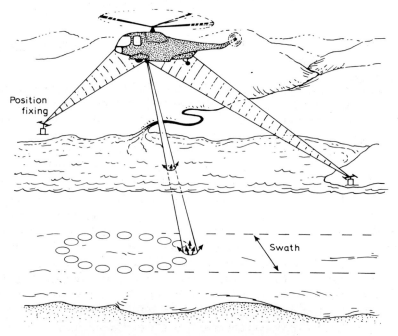

Figure 5-3 A configuration for lidar bathymetry operation (Muirhead and Cracknell, 1986)

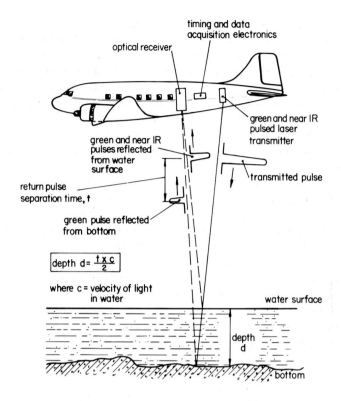

Figure 5-4 Principles of operation of a lidar bathymeter (O'Neil et al., 1980)

be determined over the whole area that is imaged in the colour photograph.

The normal flying height for this system is 1500 m. Time of day and weather are important mission considerations. To maximize depth penetration and minimize glare from the surface, a Sun angle relative to the horizon of between 18° and 25° is optimal and between 18° and 35° is acceptable. A low sea state of between 0 and 1 on the Beaufort scale is essential. Some wave action is permissible but breaking waves are not acceptable. Cloud cover should not exceed 5%. Navigational accuracy is important and special attention has to be paid to this problem over open areas of water that possess no fixed objects to assist in the identification of position. Trials of the system have been carried out successfully in one or two areas and it is hoped that the system will be introduced on an operational basis, (see Chapter 10). In the form in which we have described it the lidar bathymeter is a profiling instrument, that is it simply measures the depth at a set of points on a line vertically below the flight path of the aircraft. A further development involves the development of a scanning version of the instrument.

In many areas with high turbidity in the water, i.e. with a high concentration of suspended material, the primary problem in measuring depth with a lidar bathymeter arises from the large amount of backscattering from the water column which broadens the bottom pulse as well as producing a high "clutter" level in the region of the bottom peak. When such a situation arises there is no advantage to be gained by increasing the laser power or by range gating the receiver since the effective noise level due to this scattering increases along with the desired signal. A useful parameter for describing the performance of the sensor is the product of the mean attenuation coefficient and the maximum recorded depth.

5.3 Laser fluorosensing

Fluorescence occurs when a photon is absorbed by a target molecule and another photon is subsequently emitted with a longer wavelength. Although not all molecules fluoresce, the wavelength spectrum and the decay time spectrum of emitted photons are characteristics of the target molecules for the specific wavelength of the absorbed photons. In a remote sensing context the source of excitation photons can be either the Sun or an artificial light source. In the present context, it is the active process, involving the use of a laser as an artificial light source, which is of relevance. The remote sensing system which stimulates and analyses the fluorescence emission has become known as the laser fluorosensor.

If the target is in an aquatic environment the excitation photons may undergo Raman scattering by the water molecules. Part of the energy of the incident photons is absorbed by a vibrational energy level in the water molecule (the [OH] stretch) and so the scattered photons are shifted to a longer wavelength corresponding to f/c or $1/\lambda = 3418$ cm^{-1}. The amplitude of the Raman signal is directly proportional to the number of water molecules in the incident photon beam. This Raman line is a prominent feature of remotely-sensed fluorescence spectra taken over water and is used to estimate the depth to which the excitation photons penetrate the water. There have been a number of laser fluorosensors developed and they have been applied to the detection, identification and mapping of oil and various other materials, mostly in an aquatic environment.

Three laser fluorosensing systems, developed by the Canada Centre for Remote Sensing, by NASA and by the U.S. Environmental Protection Agency, are described in general terms by O'Neil *et al.* (1981). The system developed by the Canada Centre for Remote Sensing (O'Neil *et al.*, 1980; Zwick *et al.*, 1981) was intended primarily for the monitoring of oil pollution and was backed up by a considerable amount of work on laboratory studies of the fluorescence spectra of oils. The Airborne Oceanographic Laser (AOL) developed by NASA was designed to be used both for lidar bathymetry and for fluorosensing and has been applied to various problems. The system developed by the U.S. Environmental

Protection Agency has primarily been developed for the purpose of water-quality monitoring. This has involved the study of chlorophyll and of dissolved organic carbon (Bristow and Neilsen, 1981; Bristow *et al.*, 1981).

The generalised laser fluorosensor consists of a laser transmitter, operating in the ultraviolet part of the spectrum, an optical receiver and a data acquisition system. A laser is used, rather than any other type of light source, because it can deliver a high radiant flux density at a well defined wavelength to the target surface. In selecting the wavelength at which to operate, one endeavours to choose the wavelength at which the target molecules fluoresce most strongly. Alternatively, multiple excitation wavelengths may be used, in which case the variation of spectral shape and intensity of the resulting fluorescence emissions can be used to characterise the target. In current systems, a pulsed laser is used to allow daylight operation, target range determination and, potentially, fluorescence lifetime measurement. A block diagram of the electro-optical system of a laser fluorosensor is shown in Figure 5-5 (O'Neil *et al.*, 1980). The characteristics of the laser transmitter, including the collimator, are summarized in Table 5-1. The induced fluorescence is observed by a receiver that consists of two main subsystems, the spectrometer and the lidar altimeter. The receiver characteristics are summarized in Table 5-2. Fluorescence decay times may also be measured in the future with the addition of high-speed detectors. The telescope collects light from the point

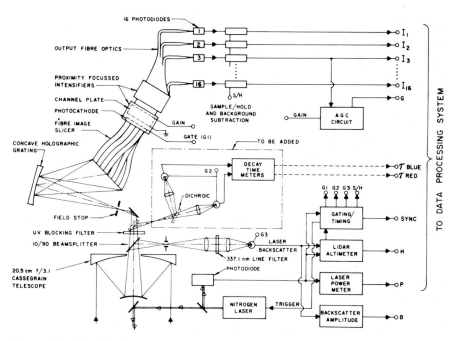

Figure 5-5 Block diagram of a fluorosensor electro-optical system (O'Neil et al.*, 1980)*

Table 5.1　Laser transmitter characteristics

Laser type	Nitrogen gas laser
Wavelength	337 nm
Pulse length	3-nsec FWHM
Pulse energy	1 mJ/pulse
Beam divergence	3 mrad × 1 mrad
Repetition rate	100 Hz

(from O'Neil et al., 1980)

Table 5.2　Laser fluorosensor receiver characteristics

Telescope	f/3·1 Dall Kirkham
Clear aperture	0·0232 m^2
Field of view	3 mrad × 1 mrad
Intensifier on-gate period	70 nsec
Nominal spectral range	386-690 nm
Nominal spectral bandpass (channels 2–15)	20 nm/channel
Noise equivalent energy*	$\sim 4 \cdot 8 \times 10^{-17}$J
Lidar altimeter range	75-750 m
Lidar altimeter resolution	1·5 m

*This is the apparent fluorescence signal (after background subtraction) collected by the receiver in one wavelength channel for a single laser pulse that equals the noise in that channel. This figure relates to the sensor performance at the time of collection of the data presented by O'Neil et al., (1980). The noise equivalent energy has been improved significantly.

(from O'Neil et al., 1980)

where the laser beam strikes the surface of the Earth. An ultraviolet blocking filter prevents backscattered laser radiation from entering the spectrometer. The visible portion of the spectrum, which includes the laser-induced fluorescence as well as the upwelling background radiance, is dispersed by a concave holographic grating and monitored by gated detectors. Gating of the detectors permits both the background solar radiance to be removed from the observed signal and the induced fluorescence emission to be measured only at a specific range from the sensor; for example, it is possible to measure the fluorescence of the surface or over a depth interval below the surface.

In the particular system considered in Figure 5-5 the received light is separated into 16 spectral channels. The first channel is centred on the water Raman line at 381 nm and is 8 nm wide. (In the figures this channel is labelled as the 380 nm channel). The spectral range from 400 nm to

660 nm is covered by 14 channels, each 20 nm wide. The 16th channel is centred at 685 nm in order to observe the chlorophyll-a fluorescence emission and is only 7 nm wide. For each laser pulse the output of each photodiode is sampled, digitized and passed to the data acquisition system, which also notes the lidar altitude, the ultraviolet backscatter amplitude, the laser pulse power and the receiver gain.

The second main subsystem in the laser fluorosensor considered in Figure 5-5 is the lidar altimeter. The lidar altimeter uses the two-way transit time of the ultraviolet laser pulse to measure the altitude of the fluorosensor above the terrain. The lidar altitude is required to gate the receiver and, along with the pulse energy and receiver gain, to normalize the fluorescence intensity and hence to estimate the fluorescence conversion efficiency of the target. Laboratory studies have shown that mineral oils fluoresce efficiently enough to be detected by a laser fluorosensor and that their fluorescence spectra not only allow oil to be distinguished from a sea-water background, but also allow classification of the oil into three groups, namely, light refined (e.g., diesel), crude, and heavy refined (e.g., bunker fuel). The fluorescence spectra of three oils typical of these groups are shown in Figure 5-6.

The fluorescence decay properties of oils have also been shown to be characteristic of the group and allow the same sort of classification to be made. It has been shown that the sea state will have a negligible effect on fluorosensor operation.

One of the principal uses of laser fluorosensing from aircraft is for oil-

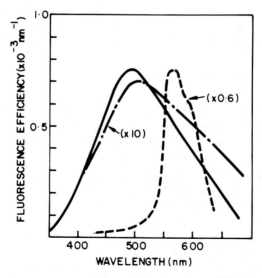

Figure 5-6 Laboratory measured fluorescence spectra of Merban crude oil (solid line), La Rosa crude oil (dash-dot line) and rhodamine WT dye (1% in water) (dashed line) (O'Neil et al., 1980)

spill detection, characterization, mapping and thickness contouring. Other uses include the mapping of chlorophyll concentrations and the mapping of fluorescent dye concentrations. When used for oil pollution surveillance, a laser fluorosensor can perform three distinct operations. The first is to detect an anomaly, the second is to identify the anomaly as oil and not some other substance, and the third is to classify the oil into the three broad categories just mentioned (Figure 5-6). Within certain limits the sensor can measure the thickness of the oil and, with systematic coverage, a slick can be mapped.

The theory of the measurement of oil film thickness in the field has been developed in conjunction with an absolute oil film fluorescence conversion efficiency measurement technique (Kung and Itzkan, 1976). The oil film thickness, d, is given by

$$d = -\left(\frac{1}{K_L + K_R}\right) \ln \left(\frac{P_{Rin}}{P_{Rout}}\right) \tag{5-1}$$

where K_L and K_R are the attenuation coefficients of the oil film at the laser and water Raman wavelengths respectively ($\lambda_L = 337$ nm and $\lambda_R = 381$ nm) and P_{Rin} and P_{Rout} are the peak detected powers in the water Raman band from beneath the oil slick and from the adjacent water mass with no slick present, respectively.

The ability of laser fluorescence to measure oil film characteristics depends on the lateral homogeneity of the water within the region of the spill, while the coefficients K_L and K_R have to be measured in the laboratory (Hoge and Kincaid, 1980). One of the largest sources of error in this thickness measurement technique is the unknown effect of the natural ageing of the oil film on the extinction coefficients, K_L and K_R. At present, no technique is known that allows for the measurement, in the field, of these extinction coefficients or of their changes with time. In addition, the Raman technique given here assumes a contiguous oil film. For oils that coagulate into small optically opaque regions (thicknesses $> 20\,\mu m$), the technique will have the effect of spatial averaging within the 3 mrad laser footprint and will yield an erroneously small thickness. Evidence from several oil spills suggests that the sensor also detects oil dispersed in the water column. The laser fluorosensor is believed to be the only sensor system able to detect oil in waters with a significant degree of ice cover.

Airborne fluorosensing systems can also be used to make measurements of chlorophyll concentration from a distance. Chlorophyll-a can be stimulated to fluoresce at a peak emission wavelength of 685 nm. Generally, fluorometers for *in situ* measurements employ an excitation wavelength of 440 nm in the blue part of the spectrum where chlorophyll-a exhibits a strong absorption band; however, the absolute fluorescence conversion efficiency depends on the algal colour group. Jarrett *et al.* (1979) have designed a fluorosensor with multiple excitation wavelengths which exploits this variation to measure relative concentrations of the various algal colour groups. The Environmental Protection Agency's laser fluorosensor is built around a pumped dye laser with a broad range of possible excitation wavelengths. On the basis of a compromise between various conflicting factors an excitation wavelength of 470 nm has been chosen for use in some remote sensing measurements. It can be shown (Bristow *et al.*, 1981) that the concentration n_f of a fluorescent material uniformly distributed through the water column is given by

$$n_f = C \frac{P_F}{P_R} \tag{5-2}$$

where P_R and P_F are the peak detected powers in the Raman band and the fluorescence band, respectively ($\lambda_R = 560$ nm, $\lambda_L = 470$ nm, $\lambda_F = 685$ nm for chlorophyll-a) and C is a constant. The constant, C, is independent of laser power and target range but contains a factor depending on the effective optical attenuation coefficient of the water column at the laser wavelength (K_L), the Raman wavelength (K_R) and the fluorescence wavelength (K_F). For equation 5-2 to be valid, the ratio $(K_L + K_R)/(K_L + K_F)$ must be constant over the survey area. For a water body where the dissolved and suspended materials change in concentration but not in character, this is believed to be a reasonable assumption. To obtain quantitative measurements of the chlorophyll concentrations with a laser fluorosensor, rather than just relative measurements, requires the results of a few *in situ* measurements of chlorophyll-a concentration made by conventional means for samples taken simultaneously from a few points under the flight path. These *in situ* measurements are needed for the calibration of the airborne data because one is not dealing with a single chemical substance but with a group of chemically-related materials, the relative concentrations of which will depend on the specific mixture of algal species present. Because the absolute fluorescence conversion efficiency depends on the recent history of photosynthetic activity of the organisms (due to changes in water temperature, salinity, and nutrient levels as well as the ambient irradiance), this calibration is essential if data are to be compared from day to day or from region to region. Similarly, an *in situ* measurement of the optical transmission of the water is needed if absolute values of the effective attenuation coefficient are to be extracted from the airborne data.

From the equation 5-2 it may be shown that the effective optical attenuation coefficient, K_V, at some intermediate wavelength, λ_V, between the laser and Raman spectral lines can be approximated by (Bristow *et al.*, 1981)

$$K_V = \frac{C_1}{P_R} \tag{5-3}$$

where C_1 is a constant. For the geometry used in an airborne laser fluorosensor, the effective optical attenuation coefficient, K_V, so obtained, is believed to be a close approximation to the beam attenuation coefficient at the intermediate wavelength.   This has been verified by *in situ* measurements.

Fluorescent dyes are often used as tracers for studying the diffusion/dispersion of, for example, sewage pollution (Valerio, 1981, 1983) and in certain aspects of hydrology (Smart and Laidlaw, 1977). The advantage of using a laser system is that, since one can use a dye that is a well-characterized chemical, one can obtain dye concentration maps without the need for extensive *in situ* sampling of the dye concentration.

In common with all optical techniques, the depth to which laser fluorosensor measurements can be made is limited by the transmission of the excitation and emission photons through the target and its environment. Any one of the materials that can be monitored by laser fluorosensing can also be monitored by grab sampling from a ship or, in some cases, by airborne and satellite multi-spectral scanners (MSS). Each method has its advantages. While *in situ* measurements or grab sample analyses are the accepted standard technique, the spatial coverage by this technique is so poor that any temporal variations over a large area are extremely difficult to unravel. For rapid large area surveys, a satellite MSS is probably the instrument of choice if there are sufficient surface measurements with which to calibrate the resulting imagery. The airborne laser fluorosensor can rapidly cover areas of moderate size and the data can be made available very quickly, with only a few surface measurements needed to extract quantitative concentrations; thus the airborne laser fluorosensor can be used to extend a very limited surface sampling grid to the dimensions suitable for the calibration of satellite imagery.

Most of the work with airborne lidar systems has been concerned with the aquatic environment; the existing demonstrations may encourage the development of new techniques for monitoring other water quality parameters directly or through their influence on fluorescent substances that can be measured directly using laser fluorosensors. Operational laser fluorosensor packages may be used in the future for routine surveillance of water quality. Airborne lidar techniques are only now being applied to land targets and it seems likely that agriculture and forestry may provide the first successful demonstrations of these techniques for land applications.

6 Ground wave and sky wave radar techniques

6.1 Introduction

Before the advent of remote sensing techniques, data on sea state and wind speeds at sea were obtained from ships and buoys and were accordingly only available for a sparse array of points. Wave heights were often simply estimated by an observer standing on the deck of a ship. Remote sensing techniques using aircraft and, more especially, satellites, have the very great advantage of being able to provide information about enormous areas of the surface of the Earth simultaneously. However, remote sensing by satellite-flown instruments using radiation from the visible or infrared parts of the electromagnetic spectrum has the serious disadvantage that the surface of the sea is often obscured by cloud. While data on wind speeds at cloud height are obtainable from a succession of satellite images, these would not necessarily be representative of wind speeds at ground level.

It is, of course, under adverse weather conditions that one is likely to be particularly anxious to obtain sea state and marine weather data. While aircraft are expensive to purchase and maintain and their use is restricted somewhat by adverse weather conditions, remote sensing techniques can provide a good deal of relevant information at low cost. Satellites are even more expensive than aircraft, though this may be forgotten if someone else has paid the large capital costs involved and only the marginal costs of the archiving and distribution of the data are paid by the user. Satellites, of course, have the advantage over other remote sensing platforms that coverage of large areas is obtained. If one is concerned with only a relatively small area of the surface of the Earth similar data can be obtained about sea state and near-surface wind speeds by using ground-based or ship-based radar systems. Figure 6-1 is taken from a review by Shearman (1981) and illustrates (though not to scale) ground wave and sky wave techniques.

The distinction should be made between imaging and non-imaging

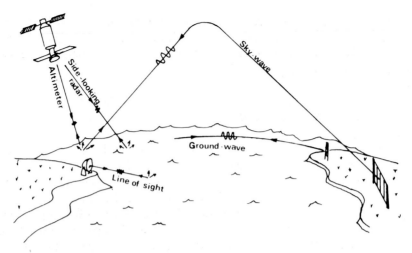

Figure 6-1 Ground and sky wave radars for oceanography (Shearman, 1981)

active microwave systems. Side-looking airborne radars (SLARs) flown on aircraft and synthetic aperture radars (SARs) flown on aircraft and on spacecraft are imaging devices and can, for instance, give information about wavelengths and about the direction of propagation of waves. A substantial computational effort involving Fourier analyses of the wave patterns is required to achieve this. In the case of synthetic aperture radar this computational effort is additional to the already quite massive computational effort involved in generating an image from the raw data (see Section 7.3). A synthetic aperture radar image is very useful for identifying ships, icebergs and iceflows. Other active microwave instruments, such as altimeters and scatterometers, do not form images but give information about wave heights and wind speeds. This information is obtained from the shapes of the return pulses received by the instruments. The altimeter (see Section 7.1) operates with short pulses travelling vertically between the instrument and the ground and is used to determine the shape of the geoid and the (rms) wave height. A scatterometer involves the use of beams that are offset from the vertical. Calibration data are used to determine wave heights and directions, and wind speeds and directions.

Among possibilities in ground-based radars for sea state studies there are three types of systems; these are, see Figure 6-1

    1. Direct line-of-sight systems;

    2. Ground wave systems; and

    3. Sky wave systems.

The first of these uses conventional microwave frequencies while the second and third use longer wavelength radio waves, decametric waves, which correspond to conventional medium-wave broadcast band frequencies. Microwave radar is limited to use within the direct line-of-sight and cannot be used to see beyond the horizon. A radar mounted on a cliff is unlikely to exceed a distance of 30 – 50 km. Microwave radar systems are discussed in Chapter 7.

6.2 Ground wave systems

The origin of the use of decametric radar for the study of sea state dates from the work of Crombie (1955) who discovered that with radio waves of frequency of 13.56 MHz, that is 22 m wavelength, the radar echo from the sea detected at a coastal site had a characteristic Doppler spectrum with one strongly dominant frequency component. The frequency of this component was shifted by 0.376 Hz which corresponded to the Doppler shift expected from the velocity of the sea waves with a wavelength equal to half the wavelength of the radio wave travelling towards the radar. This means that the radio waves interact with sea waves of comparable wavelength in a resonant fashion that is analogous to the Bragg scattering of X-rays by the rows of atoms in a crystal. Crombie envisaged a coastal-based radar system, using multi-frequency radars with steerable beams, to provide a radar spectrometer for studying waves on the surface of the sea. Such radars would have greater range than the direct line-of-sight microwave radars erected on coastal sites (see Figure 6-1) because they would be operating at longer wavelengths, namely tens of metres. These waves, which are referred to as "ground waves", bend around the surface of the Earth so that such a ground wave radar would be expected to have a range of between 100 km and 500 km depending on the power and frequency of the radar used.

If the radio waves strike the sea surface at an angle, say Δ, the Bragg scattering condition is $2L\cos\Delta = \lambda$, where L is the sea-surface wavelength and λ is the radio wavelength. For ground waves the radio waves strike the sea surface at grazing incidence and the Bragg scattering condition simplifies to $2L = \lambda$. There is not, of course, a single wavelength alone present in the waves on the surface of the sea; there is a complicated pattern of waves with a wind-dependent spectrum of wavelengths and spread of directions. The importance of the Bragg-scattering mechanism is that the radar can be used to study a particular selected wavelength component in the chaotic pattern of waves on the sea surface. Quantities readily derivable from the Bragg resonant lines are the wind direction, from the relative magnitude of the approach and recede amplitude (see Figure 6-2a), and the radial component of the current (see Figure 6-2c). The possibility of determining the current directly from the Doppler shift does not arise with a synthetic aperture radar because the Doppler shift associated with the moving target cannot easily be separated from the Doppler shift associated with the movement of the aircraft or satellite platform that carries the radar.

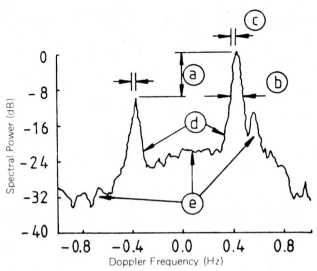

Figure 6-2 Features of radar spectra used for sea-state measurement: (a) ratio of two first-order Bragg lines — wind direction; (b) −10 dB width of larger first-order Bragg line — wind speed; (c) Doppler shift of first-order Bragg lines from expected values — radial component of surface current; (d) magnitudes of first-order Bragg lines — ocean wave height spectrum for one wave-frequency and direction; and (e) magnitude of second-order structure — ocean wave height spectrum for all wave-frequencies and directions (sky wave data for 10.00 UT, 23 August 1978, frequency 15 MHz data-window, Hanning, FFT 1024 points, averages 10, slant range 1125 km) (Shearman, 1981)

It had originally been supposed that the mapping of currents using ground wave radar would require the use of a directional antenna array to resolve the sea areas in azimuth. However, following further work by Crombie, a new approach has been adopted which had been incorporated in the NOAA "CODAR" current mapping radar (Barrick *et al.*, 1977). This involves using a broad-beam transmitter at a high frequency (~26 MHz). The returning radio echoes are received separately on four whip antennae located at the corners of a square. A Doppler spectrum is determined from the signals received at each of the four whip antennae and the phases of the components of a particular Doppler shift in each of the spectra are then compared to deduce the azimuthal direction from which that component has come. With two such radars on two separate sites the radial components of the currents can be determined, with reference to each site, and the two sets of results can then be combined to yield the current as a vector field.

6.3 Sky wave systems

The "sky wave" radar, see Figure 6-1, involves decametric waves that are reflected by the ionosphere and consequently follow the curvature of the Earth in a manner that is very familiar to short-wave radio listeners. These waves are able to cover very large distances around the Earth. Sky wave radars can be used accordingly to study sea-surface waves at distances between 1000 km and 3000 km from the radar installation. The observation of data on sea-state spectra gathered by sky wave radar was first reported by Ward (1969). As with the ground-wave spectra, sky wave radar depends on the selectivity of wavelengths achieved by Bragg scattering at the surface of the sea. There is, however, a difference between the Bragg scattering of ground waves and of sky waves. In the case of ground waves the radio waves strike the sea surface at grazing incidence, but in the case of the sky waves the radio waves strike the sea surface obliquely, say at an angle Δ, and the full Bragg condition $2L\cos\Delta = \lambda$ applies.

For the waves on the surface of the sea it is only the components of the wave vector directly towards or away from the radar that are involved in the Bragg condition. The relative amplitudes of the positively and negatively Doppler-shifted lines in the spectrum of the radar echo from a particular area of the sea indicate the ratio of the energy in approaching and receding wind-driven sea waves. Should there be only a positively shifted line present the wind is blowing directly towards the radar; conversely, should there be only a negatively shifted line the wind is blowing directly away from the radar. If the polar diagram of the wind-driven waves about the mean wind direction is known, the measured ratio of the positive and negative Doppler shifts enables the direction of the mean wind to be deduced. This is achieved by rotating the wave-energy polar diagram relative to the direction of the radar beam until the radar beam's direction cuts the polar diagram with the correct ratio, see Figure 6-3. There will be two wind directions that satisfy this condition; these directions will be symmetrically oriented on the left and right of the direction of the radar beam. This ambiguity can be resolved using observations from a sector of radar beam directions and making use of the continuity conditions for wind circulation, see Figure 6-4.

In practice it is observed that the positive and negative Doppler shifts are not quite equal in magnitude. This arises because there is an extra Doppler shift arising from the bodily movement of the water surface on which the waves travel, this movement being the surface current. The radar, however, is only capable of determining the component of the total surface current along the direction of the radar beam.

Figure 6-2 shows a sky wave radar spectrum labelled with the various oceanographic and meteorological quantities which can be derived from it. As well as the quantities that have already been mentioned, there are other quantities which can be derived from the second-order features. It should be noted that current measurements from sky wave radars are contaminated by extra Doppler shifts due to ionospheric layer height changes. If current measurements are to be attempted it is necessary to calibrate this

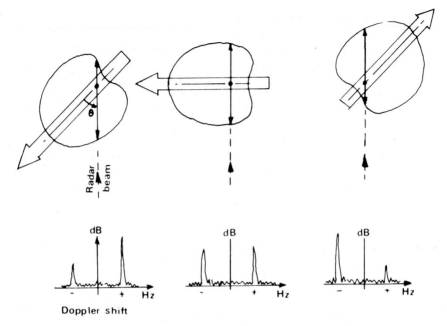

Figure 6-3 Typical spectra obtained for different wind orientations relative to the radar boresight (Shearman, 1981)

ionospheric Doppler shift. This may be done by considering the echoes from an island for instance.

The simple ideas of Bragg scattering which we have mentioned are valuable in identifying the particular wavelength of radio wave that will be selected to contribute to the return pulse, but it does not give us a value for the actual intensity of the backscattered radio waves nor does it take into account second-order effects. This can be tackled by the extension and adaptation to electromagnetic scattering given by Rice (1951) and Barrick (1971, 1972*a*, 1972*b*, 1977*a*, 1977*b*) and Barrick and Weber (1977) of the treatment, originally due to Lord Rayleigh, of the scattering of sound from a corrugated surface. This is essentially a perturbation theory argument. A plane radio wave is considered to be incident on a corrugated or rough conducting surface and the vector sum of the incident and scattered waves at the surface of the conductor must satisfy the boundary conditions on the electromagnetic fields, in particular that the tangential component of the electric field is zero. More complicated boundary conditions apply if one takes into account the fact that the sea water is not a perfect conductor and that its relative permittivity is not exactly equal to unity.

The resultant electric field of all the scattered waves will have a component parallel to the surface of the water which must cancel out exactly with the component of the incident wave parallel to the surface. The scattering problem therefore involves the determination of the phases, amplitudes and polarisations of the scattered waves that will satisfy this condition. Consider a plane radio wave with wavelength λ_0 incident with grazing angle Δ_i on a sea surface with a sinusoidal swell wave of height H ($H \ll \lambda_0$) and wavelength L travelling with its velocity in the plane of incidence, see Figure 6-5(a). There will be three scattered waves, with grazing angles of reflection of Δ_i, Δ_{s^+} and Δ_{s^-}, where

$$\cos \Delta_{s\pm} = \cos \Delta_i \pm \frac{\lambda_0}{L} \qquad (6-1)$$

If the condition that one of these scattered waves is to return along the direction of the incident wave is imposed, then $\Delta_{s^-} = \pi - \Delta_i$ so that

$$\frac{\lambda_0}{L} = \cos \Delta_i - \cos \Delta_{s^-}$$

$$= \cos \Delta_i - \cos (\pi - \Delta_i)$$

$$= 2 \cos \Delta_i$$

i.e.
$$\lambda_0 = 2L \cos \Delta_i \qquad (6-2)$$

which is just the Bragg condition.

If the condition that $H \ll \lambda_0$ is relaxed then the scattered-wave spectrum will contain additional scattered waves with grazing angles given by

$$\cos \Delta_s = \cos \Delta_i \pm \frac{\lambda_0}{L} \qquad (6-3)$$

However, it has been shown that for decametric radio waves incident on the surface of the sea this higher-order scattering is unimportant in most cases; it only becomes important for very short radio wavelengths and for very high sea states (Barrick, 1972b).

The condition expressed in equation 6-1 can be regarded as one component of a vector equation

$$\mathbf{k}_s = \mathbf{k}_i \pm \mathbf{K} \qquad (6-4)$$

where $\mathbf{k}_i$, $\mathbf{k}_s$ and $\mathbf{K}$ are vectors in the horizontal plane and associated with the incident and reflected radio waves and the swell, respectively.

The above discussion supposes that the incident radio wave, the normal and the wave vector of the swell are in the same plane. This can be generalized to cover the case when the swell waves are travelling in a direction that is not in the plane of incidence of the radio wave, see Figure 6-5(c).

The relationship between the Doppler shift observed in the scattered radio wave and the swell wave is

$$f_s = f_i \pm f_w \qquad (6-5)$$

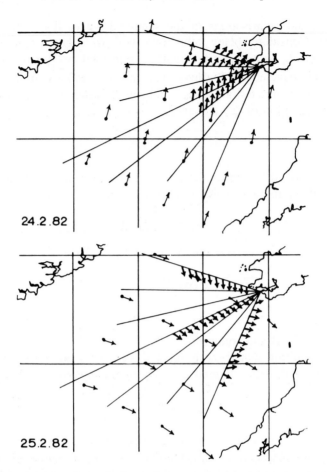

Figure 6-4 Radar-deduced wind directions (heavy arrows) compared with Meteorological-Office analysed winds. The discrepancies in the lower picture are due to the multiple peak structure on this bearing (Wyatt, 1983)

where f_s is the frequency of the scattered wave, f_i is the frequency of the incident wave and f_w is the frequency of the water wave.

An attempt could be made to determine the wave directional spectrum by using first-order returns and by using a range of radar frequencies and radar look directions. This would involve a complicated hardware system. In practice it is likely to be easier to retain a relatively simple hardware system and to use second-order, or multiple scattering, effects to provide wave directional spectrum data. It is important to understand and quantify the second-order effects because of the opportunities they provide for measuring the wave height, the non-directional wave-height spectrum and the directional wave height spectrum by inversion processes (see, for example, Shearman, 1981). The arguments given above can be extended to

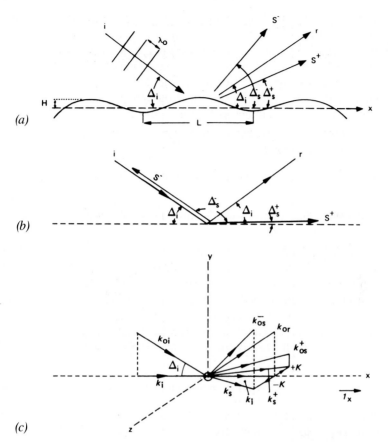

Figure 6-5 (a) Scattering from a sinusoidally corrugated surface with $H < \lambda_0$. i, r and s indicate the incident, specularly reflected and first-order scattered waves; (b) The back-scatter case, $\Delta_s = \pi - \Delta_i$; (c) The case showing general three-dimensional geometry with vector construction for scattered radio waves

multiple scattering. For successive scattering from two water waves with wave vectors $\mathbf{K}_1$ and $\mathbf{K}_2$ we would have in place of equations 6-4 and 6-5

$$\mathbf{k}_s = \mathbf{k}_i \pm \mathbf{K}_1 \pm \mathbf{K}_2 \tag{6-6}$$

and

$$f_s = f_i \pm f_{w_1} \pm f_{w_2} \tag{6-7}$$

where f_{w_1} and f_{w_2} are the frequencies of the two waves.

If we impose the additonal constraint that the scattered radio wave must constitute a freely propagating wave of velocity c, then

$$f_s \lambda = \frac{2\pi f_s}{k_s} = c \tag{6-8}$$

This results in scattering from two sea waves travelling at right angles, analogous to the corner reflector in optics or microwave radar.

The method used by Barrick involves using a Fourier series expansion of the sea-surface height and a Fourier series expansion of the electromagnetic field. The electromagnetic fields at the boundary, and hence the coefficients in the expansion of the electromagnetic fields, are expanded using perturbation theory subject to the following conditions:

1. The height of the waves must be very small compared with the radio wavelength;

2. The slopes at the surface must be small compared with unity; and

3. The impedance of the surface must be small compared with the impedance of free space.

First-order in the perturbation series corresponds to the simple Bragg scattering we have described, while second-order corresponds to the "corner-reflector" scattering by two waves. A perturbation series expansion of the Fourier coefficients used in the description of the sea surface is also used and an expression for the second-order scattered electromagnetic field due to the second-order sea-surface wave field can be obtained. In the notation used by Wyatt (1983) the backscattering cross section takes the form

$$\sigma(\omega) = \sigma^1(\omega) + \sigma^2(\omega) \qquad (6\text{-}9)$$

$\sigma^1(\omega)$ and $\sigma^2(\omega)$ are the first-order and second-order scattering cross sections, respectively, and they are given by

$$\sigma^1(\omega) = 2^6 \pi k_0^4 \sum_{m=\pm 1} S(-2m\mathbf{k}_0)\delta(\omega - m\omega_B) \qquad (6\text{-}10)$$

$\mathbf{k}_0$ = radar wavenumber
ω = Doppler frequency
$\omega_B = \sqrt{2gk_0}$ = Bragg resonant frequency
$S(\mathbf{k})$ = sea-wave directional spectrum

$$\sigma^2(\omega) = 2^6 \pi k_0^4 \sum_{m,m'=\pm 1} \int_0^\infty \int_{-\pi}^\pi |\Gamma|^2 S(m\mathbf{k})S(m'\mathbf{k}')$$

$$\times \ \delta(\omega - m\sqrt{gk} - m'\sqrt{gk'})k \ dk \ d\theta \qquad (6\text{-}11)$$

where $\mathbf{k}, \mathbf{k}'$ = wavenumbers of two interacting waves where
$\qquad \mathbf{k} + \mathbf{k}' = -2\mathbf{k}_0$
$\quad k,\theta$ = polar coordinates of $\mathbf{k}$
$\quad \Gamma$ = coupling coefficient = $\Gamma_H + \Gamma_{EM}$
$\quad \Gamma_H$ = hydrodynamic coupling coefficient

$$= -\frac{i}{2} \frac{k + k' - (kk' - \mathbf{k} \cdot \mathbf{k}')}{mm'\sqrt{kk'}} \frac{(\omega^2 + \omega_B{}^2)}{(\omega^2 - \omega_B{}^2)} \tag{6-12}$$

Γ_{EM} = electromagnetic coupling coefficient

$$= \frac{1}{2} \frac{(\mathbf{k} \cdot \mathbf{k}_0)(\mathbf{k}' \cdot \mathbf{k}_0)/(k_0^2 - 2\mathbf{k} \cdot \mathbf{k}')}{\sqrt{k \cdot k'} + k_0 \Delta} \tag{6-13}$$

Δ = normalized electrical impedance of the sea surface.

The problem then is to invert equations 6-10 and 6-11 to determine $S(\mathbf{k})$, the sea-wave directional spectrum, from the measured backscattering cross section. These equations enable one to compute the Doppler spectrum, i.e. the power of the radio wave echo as a function of Doppler frequency, by both first-order and second-order mechanisms, given the sea-wave height spectrum in terms of wave vector. However, no simple direct technique is available for the inversion of equations 6-10 and 6-11 to obtain the sea-wave spectrum from the measured radar returns. One approach is to simplify the equation and thereby obtain a solution for a restricted set of conditions. Alternatively, a model can be assumed for $S(\mathbf{k})$ including some parameters, in order to calculate $\sigma^1(\omega)$ and to determine the values of the parameters by fitting to the measured values of $\sigma^2(\omega)$; for further details see Wyatt (1983).

7 Active microwave instruments

Three important active microwave instruments, the altimeter, the scatterometer and the synthetic aperture radar, are considered in this chapter. Examples of each of these have been flown on aircraft and an example of each was flown on the SEASAT satellite.

7.1 The altimeter

The SEASAT altimeter was designed in response to a requirement for accurate determination of the Earth's geoid, that is the long-term mean equilibrium sea surface. This requires

1. a very accurate measurement of the distance from the satellite to the surface of the sea vertically below; and

2. a very accurate knowledge of the orbit of the satellite.

Other altimeters had previously been flown on aircraft and satellites leading up to the SEASAT mission. The principal measurement made by an altimeter is of the time taken for the round trip of a very short pulse of microwave energy that is transmitted vertically downward by the satellite, reflected at the surface of the Earth (sea or land) and then received back again at the satellite. The distance of the satellite above the surface of the sea is then given by

$$h = \tfrac{1}{2}ct \qquad (7\text{-}1)$$

where c is the speed of light, (see Figure 7-1).

The SEASAT altimeter transmitted short pulses at 13.5 GHz with duration of 3.2 µs and pulse repetition rate of 1020 Hz using an antenna of diameter 1 m looking vertically downward. The altimeter was designed to achieve an accuracy of ±10 cm in the determination of the geoid (details of the design of the instrument are given by Townsend, 1980). In order to achieve this accuracy it is necessary to determine the distance between the

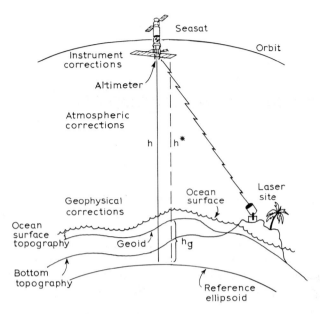

Figure 7-1 Schematic of SEASAT data collection, modelling and tracking system

surface of the sea and the satellite with considerably more accuracy than this figure. The altitude h^* measured with respect to a reference ellipsoid, see Figure 7-1, can be expressed as

$$h^* = h + h_{sg} + h_i + h_a + h_s + h_g + h_t + h_0 + e \qquad (7\text{-}2)$$

where h^* is the distance from the centre of mass of the satellite to the reference ellipsoid at the sub-satellite point. h is the height of the satellite above the surface of the sea as measured by the altimeter; h_{sg} represents the effects of spacecraft geometry, including the distance from the altimeter feed to the centre of mass of the satellite, and the effect of not pointing vertically; h_i is the total of all instrument delays and residual biases; h_a is the total atmospheric correction; h_s is the correction due to the surface and radar pulse interaction and skewness in the surface wave height distributions; h_g is the sub-satellite geoid height; h_t is the correction for solid earth and ocean tides; h_0 is the ocean-surface topography due to ocean circulations, barometric effects, wind pile-up, etc.; and e represents random measurement errors.

What is actually measured is h and this is measured to an accuracy of about ± 5 cm. Prior to the advent of SEASAT the geoid was only known to an accuracy of about ± 1 m; the idea is to measure or to calculate all the other quantities in equation 7-2 with sufficient accuracy that this equation can be used to determine the height of the geoid to an accuracy of ± 10 cm. Some examples of results obtained for height measurements with the SEASAT altimeter are shown in Figures 7-2 to 7-4.

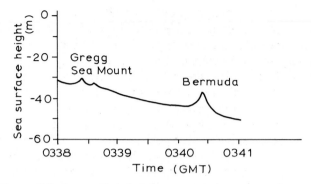

Figure 7-2 Sea surface height over sea mount-type features

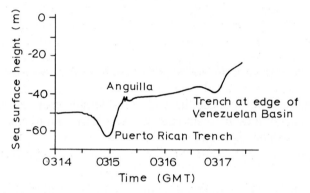

Figure 7-3 Sea surface height over trench-type features

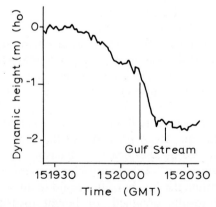

Figure 7-4 Dynamic height over the Gulf Stream

In addition to determining the range from the time of flight for the pulse, one can also extract information from the shape of the return pulse. For a perfectly flat horizontal sea surface the leading edge of the return pulse would be a very sharp square step function corresponding to a time given by equation 7-1 for radiation that travels vertically; radiation travelling at an angle inclined to the vertical will arrive slightly later. However, if there are large waves present on the surface of the sea there will be some radiation reflected from the tops of the waves, corresponding to a slightly smaller value of h and therefore a slightly smaller value of t; in the same way, there will be an extra delay experienced by radiation reflected by the troughs of the waves. Thus for a rough sea the leading edge of the return pulse will become considerably less sharp, (see Figure 7-5).

The size of the "footprint" of a radar altimeter pulse on the water surface, that is the area of the surface of the sea that contributes to the return pulse received by the altimeter, also depends on the sea state. At a given distance from nadir the probability of a wave having facets which reflect radiation back to the satellite increases with increasing roughness of the surface of the sea. The area actually illuminated is the same; it is the area from which reflected radiation is returned to the satellite that varies with sea state. For a low sea state the spot size for the SEASAT altimeter was about 1.6 km. For higher sea state the spot size increased up to about 12 km. The ability to measure the change in the shape of the leading edge of the received pulse provides an estimate of the sea state, i.e., of the significant wave height, SWH or $H_{1/3}$. The significant wave height is defined as the average height of the highest one third of all the waves and is usually taken to be four times the rms wave height (Longuet-Higgins, 1952).

The level of backscatter from an object is usually expressed as the radar cross section (RCS or σ) which is defined as the area intercepting that amount of power which, when scattered isotropically, produces an echo equal to that of the object. For an extended target, such as the sea surface, the backscatter is expressed as the normalised radar cross-section (NRCS or

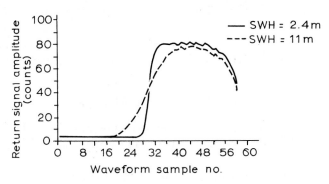

Figure 7-5 Return pulse shape as a function of SWH (significant wave height)

σ'), which is the RCS per unit area, given by

$$\sigma^0 = \frac{64\pi^3 R^4}{\lambda^2 L_S G_0^2 \left(\dfrac{G}{G_0}\right)^2 A} \frac{P_R}{P_T} \tag{7-3}$$

where R is the slant range to the target, A is the area of the target, L_S includes atmospheric attenuation and other system losses, G_0 is the peak antenna gain, G/G_0 is the peak antenna gain in the target direction, P_T is the transmitted power, and P_R is the received power. Equation 7-3 is known as the radar equation. The value of the radar cross section depends on wind stress at the surface.

From an analysis of the return pulse received at an altimeter one can obtain a value for σ_N°, the normal incidence backscatter coefficient; from this in turn one can obtain the wind speed, but not the direction of the wind. In planning the SEASAT mission the objectives set, in terms of accuracy, were:

- Height measurements ± 10 cm;

- $H_{1/3}$ (in the range 1-20 m), ± 0.5 m or $\pm 10\%$ (whichever is larger);

- Wind speed ± 2 ms^{-1}; $\sigma_N^{\circ} \pm 1$ dB.

The determination of the wind speed from the shape of the return pulse, via the significant wave height, is not done by using an exact theoretical formula. The relationship between the change in shape of the return pulse and the value of $H_{1/3}$ at the surface was determined empirically beforehand and in processing the altimeter data from the satellite a look-up table containing this empirical relationship was used. A comparison between the results obtained from the SEASAT altimeter and from buoy measurements is presented in Figure 7-6. Figure 7-7 shows the results obtained for $H_{1/3}$ from the SEASAT altimeter for an orbit which passed very close to a hurricane, Hurricane Fico, on 16 July 1978. In this data values of $H_{1/3}$ up to 10 m were obtained.

The determination of the wind speed is carried out via the backscatter coefficient, σ_N°. The value of σ_N° is determined from measurements made on the received signal, where

$$\sigma_N^{\circ} = a_0 + a_1(AGC) + a_2(h) + L_P + L_a \tag{7-4}$$

where AGC is the automatic gain control attenuation, h is the measured height, L_p represents off-nadir pointing losses and L_a represents atmospheric attenuation.

For SEASAT the values of a_0, $a_1(AGC)$ and $a_2(h)$ were determined from pre-launch testing and by comparison with the GEOS-3 satellite altimeter at

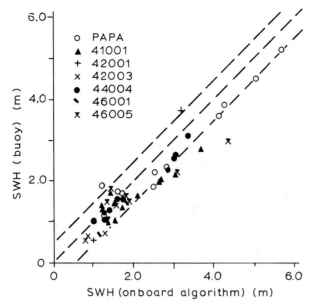

Figure 7-6 Scatter diagram comparing significant wave height estimates from the NOAA buoy network and ocean station PAPA with SEASAT altimeter on-board processor estimates (51 observations)

points where the orbits of the two satellites intersected. The calibration curve used to convert $\sigma_N°$ into windspeed was obtained using the altimeter on the GEOS-3 satellite which had been calibrated with *in situ* data obtained from data buoys equipped to determine windspeeds. Comparisons

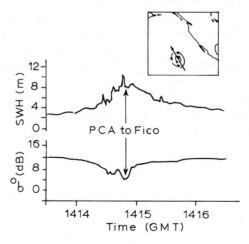

Figure 7-7 Altimeter measurements over Hurricane Fico

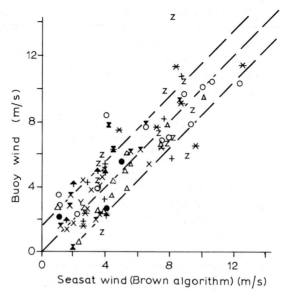

Figure 7-8 A scatter plot of SEASAT radar altimeter inferred wind speeds as a function of the corresponding buoy measurements (Guymer, 1987)

between windspeeds derived from the SEASAT altimeter and from *in situ* measurements using data buoys are shown in Figure 7-8.

7.2 The scatterometer

The altimeter, which has just been described in Section 7.1, uses just one beam directed vertically downwards from the spacecraft and enables the speed of the wind to be determined to $\pm 2 \text{ ms}^{-1}$, although the direction of the wind cannot be determined. The scatterometer uses a more complicated arrangement that actually uses four radar beams and enables the direction as well as the speed of the wind to be determined. As soon as radar had been invented it was found that, at low elevation angles, surrounding objects and terrain caused large echoes and often obliterated genuine targets; this is the well-known phenomenon of clutter. Under usual circumstances, of course, the aim is to reduce this clutter. However, research on the clutter phenomenon showed that the backscattered echo became larger with increasing wind speed. This led to the idea of using the clutter, or backscattering, to measure wind speed remotely (Jones and Schroeder, 1978; Ross and Jones, 1978; Offiler, 1983).

The backscattering arises when the Bragg condition, which involves constructive interference between reflections from successive waves on the sea surface, is satisfied

$$\lambda_s \sin \theta_i = \tfrac{1}{2} n \lambda \qquad (7\text{-}5)$$

where λ_s is the wavelength on the surface, λ is the microwave wavelength, θ_i is the angle of incidence (measured from the vertical) and n is a small integer. For typical microwave radiation the value of λ is about 2 or 3 cm, for the lowest order of reflection $n=1$, and so λ_s must also be of the order of a few centimetres. Thus the reflections arise from the capillary waves superimposed on the much longer wavelength gravity waves.

The problem of determining the wind speed from the radar backscattering cross section has already been mentioned in Section 7.1 regarding the altimeter. The difficulty is to establish the detailed relationship between windspeed and backscattering cross section. A similar problem exists with the extraction of wind velocities from scatterometer data. The relationship between radar backscattering cross section and wind velocity has been established empirically; it was not been determined theoretically from first principles. This has been done using experimental data from wind-wave tanks and also by calibrating scatterometers on fixed platforms, on aircraft and on satellites with the aid of simultaneous *in situ* data gathered at the ocean surface. The backscattering cross section $\sigma°$ increases with increasing wind speed, decreases with increasing angle of incidence and depends on the beam azimuth angle relative to the wind direction (Schroeder *et al.*, 1982); it is generally lower for horizontal polarization than for vertical polarization and it appears to depend very little on the microwave frequency in the range 5 – 20 GHz. An empirical formula that is used for the backscattering coefficient is

$$\sigma^2 = a_0(U,\theta_i,P) + a_1(U,P) \cos \phi + a_2(U,P) \cos 2\phi \qquad (7\text{-}6)$$

or

$$\sigma^2 = G(\phi,\theta_i,P) + H(\phi,\theta_i,P) \log_{10}U \qquad (7\text{-}7)$$

where U is the wind speed, ϕ is the relative wind direction, and P indicates whether the polarization is vertical or horizontal. The coefficients a_0, a_1 and a_3, which are proportional to $\log_{10}U$, or the functions $G(\phi,\theta_i,P)$ and $H(\phi,\theta_i,P)$ are derived from fitting measured backscattering results with known wind speeds and directions in calibration experiments. The form of the backscattering coefficient as a function of wind speed and direction is shown in Figure 7-9. Originally these functions were determined with data from scatterometers flown on aircraft but following the launch of SEASAT the values of these functions have been refined further. Assuming that the functions in equation 7-6 have been determined, one can then use this equation with measurements of $\sigma°$ for two or more azimuth angles ϕ to determine both the wind speed and the wind direction.

The scatterometer on the SEASAT satellite used four beams altogether; two of them were pointing forward, at 45° to the direction of flight of the satellite, and two of them were pointing aft, also at 45° to the direction of

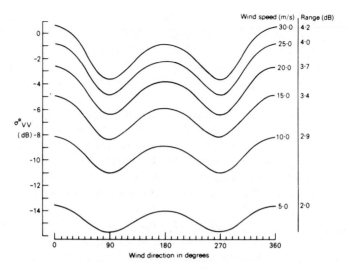

Figure 7-9 Backscatter cross section, σ°, against relative wind direction for various wind speeds. Vertical polarisation of 30° incidence angle (Offiler, 1983)

flight, see Figure 7-10. Two looks at a given area on the surface of the sea were obtained from the forward-pointing and aft-pointing beams on one side of the spacecraft; the change, as a result of Earth rotation, in the area of the surface actually viewed is quite small. The half-power beamwidths were 0.5° in the horizontal plane and about 25° in the vertical plane. This gave a swath width of about 500 km on each side, going from 200 km to 700 km away from the sub-satellite track. The return signals were separated to give backscattering data from 12 areas, or cells, along the strip of sea surface being illuminated by the transmitted pulse. The spatial resolution is thus approximately 50 km. The extraction of the wind speed and direction from the satellite data involves the following steps:

1. Identifying the position of each cell on the surface of the Earth and determining the area of the cell and the slant range;

2. Calculating the ratio of the received power to the transmitted power;

3. Determining the values of the system losses and the antenna gain in the cell direction from the pre-flight calibration data;

4. Calculating σ° from the radar equation and correcting this for atmospheric attenuation derived from the SMMR (Scanning Multi-channel Microwave Radiometer) (also on the SEASAT satellite) as well as for other instrumental biases.

It is then necessary to combine the data from the two views of a given cell from the fore and aft beams and thence determine the wind speed and

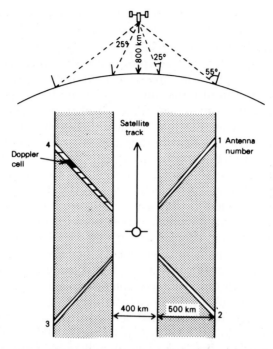

Figure 7-10 The SEASAT scatterometer viewing geometry: section in the plane of beams 1 and 3 (top diagram), beam illumination pattern and ground swath (bottom diagram). This scatterometer operated at 14·6 GHz (K$_u$-band) and a set of Doppler filters defined 15 cells in each antenna beam. Either horizontal or vertical polarisation measurements of backscatter could be made

direction; this is done using look-up tables for the functions $G(\phi,\theta_i,P)$ and $H(\phi,\theta_i,P)$. The answer, however, is not necessarily unique; there can be up to four solutions, each with similar values of the wind speed but with quite different directions, see Figure 7-9. The scatterometer on SEASAT was designed to measure the surface wind velocity with an accuracy of 2 ms^{-1} or 10% (whichever is the greater) in the speed and 20° in direction.

In spite of the relatively short duration of the SEASAT satellite before it ceased to function, there were some checks done to evaluate the derived parameters by comparison with *in situ* measurements over certain test areas. One such exercise was the Gulf of Alaska Experiment (GOASEX) which involved several oceanographic research vessels and buoys and an aircraft carrying a scatterometer similar to that flown on SEASAT (Jones *et al.*, 1979). Comparisons with the results from the *in situ* measurements showed that the results obtained from the SEASAT scatterometer were generally correct to the level of accuracy specified at the design stage, although systematic errors were detected and this information was used to update the algorithms used for processing the satellite data (Schroeder *et al.*, 1982). A second example was the Joint Air-Sea Interaction (JASIN) project which took place in the north Atlantic between Scotland and Iceland during the period that SEASAT was operational. Results from the JASIN project also showed that the wind vectors derived from the SEASAT scatterometer data were accurate well within the values specified at the design stage. Again these results were used to refine further the algorithms used to derive the wind vectors from the scatterometer data for other areas (Jones *et al.*, 1981; Offiler, 1983). The Satellite Meteorology Branch of the U.K. Meteorological Office has made a thorough investigation of the SEASAT scatterometer wind measurements, using data for the JASIN area, which covers a period of two months. The Institute of Oceanographic Sciences (Wormley, U.K.) had collated much of the JASIN data, but the wind data measured *in situ* applied to the actual height of the anemometer that provided the measurements which varied from 2.5 m above sea level to 23 m above sea level. For comparison with the SEASAT data the wind data was corrected to a common height of 19.5 m above sea level. Each SEASAT scatterometer value of the wind velocity was then paired, if possible, with a JASIN observation within 60 km and 30 minutes; a total of 2724 such pairs were obtained. Since there may be more than one possible solution for the direction of the wind derived from the scatterometer data, the value that was closest in direction to the JASIN value was chosen. Comparisons between the wind speeds obtained from the SEASAT scatterometer and the JASIN surface data are shown in Figure 7-11(a); similar comparisons for the direction are given in Figure 7-11(b). Overall, the scatterometer-derived wind velocities agreed with the surface data to within $\pm 1.7\,\mathrm{ms}^{-1}$ in speed and $\pm 17°$ in direction. However, there is evidence from data from one particular SEASAT orbit that serious errors in scatterometer-derived windspeeds may be obtained when thunderstorms are present (Guymer *et al.*, 1981; Offiler, 1983).

One of the special advantages of satellite-derived data is the high spatial density of data collected. This is illustrated rather well by the example of a cold front shown in Figure 7-12(a); and Figure 7-13. Figure 7-12(a) shows the synoptic situation at midnight GMT on 31 August 1978 and Figure 7-12(b) shows the wind field; these were both derived from the Meteorological Office's 10-level model objective analysis on a 100 km grid. Fronts have been added manually, by subjective analysis. The low pressure over Iceland had been moving north-eastwards, bringing its associated fronts over the JASIN area by midnight. On 31 August 1978 SEASAT passed just south of Iceland at 0050 GMT enabling the scatterometer to measure winds in this area, see Figure 7-13 which also shows the observations and analysis at 0100 GMT. The points *M* and *T* indicate two stations which happened to be on either side of the cold front. At most points there are four possible solutions indicated, but the front itself shows clearly in the scatterometer-derived winds as a line of points at which there are only two, rather than four, solutions. With experience, synoptic

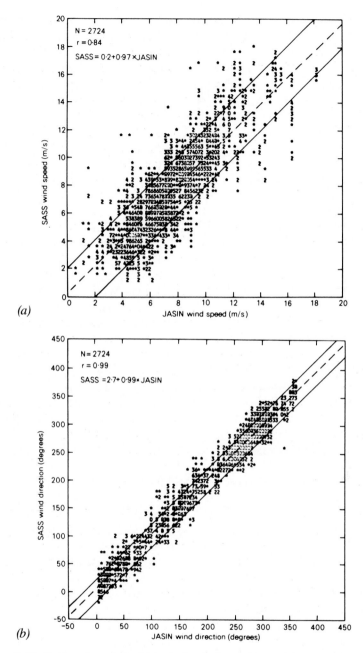

Figure 7-11 Scatter diagrams of (a) wind speed and (b) wind direction measurements made by the SEASAT scatterometer against co-located JASIN observations. The design rms limits of 2 ms⁻¹ and 20° are indicated by the solid parallel lines and the least-squares regression fit by the dashed line. Key: *=1 observation pair; 2=2 coincident observations; etc., '0'=10 and '@'=more than 10 (Offiler, 1983)

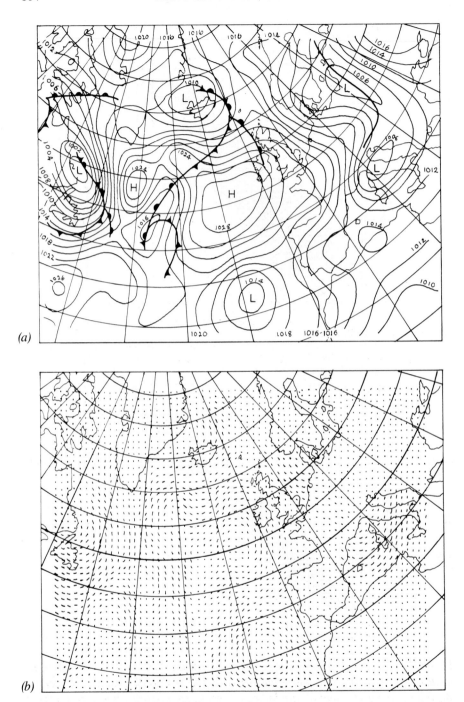

Figure 7-12 Example of (a) mean sea level pressure, and (b) 1000 mbar vector winds for 31 August 1978

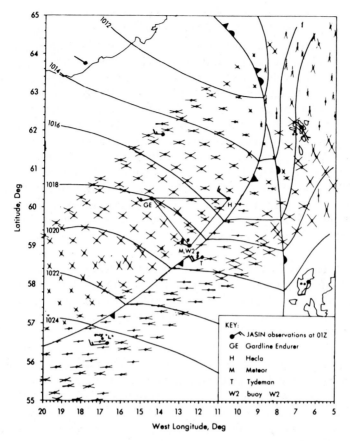

Figure 7-13 Cold front example, 0050 GMT, orbit 930 (vertical polarisation) (Offiler, 1983)

features such as fronts, and especially low pressure centres, can be positioned accurately, even with the present level of ambiguity of solutions. The subjective analysis of scatterometer-derived wind fields has been successfully demonstrated by, for example, Wurtele *et al.* (1982).

7.3 Synthetic aperture radar

Conventional remote sensing of the surface of the Earth from aircraft or spacecraft involves using either cameras or scanners which produce images in a rather direct manner. These instruments are passive instruments; they receive the radiation which happens to fall upon them and select the particular range of wavelengths that have been chosen for the instrument. When these instruments are operating at visible or infrared wavelengths they are capable of quite good spatial resolution (80 m for LANDSAT

MSS, 30 m for the Thematic Mapper, 20 m for SPOT operating in its multi-spectral mode and 10 m for SPOT operating in its panchromatic mode). However, at visible and infrared wavelengths these instruments are not able to see through clouds and, if it is cloudy, they produce images of the top of the clouds and not the surface of the Earth. By moving into the microwave part of the electromagnetic spectrum it is possible to see through clouds and hence to obtain images of the surface of the Earth even when the weather is cloudy, provided that there is not too much precipitation. Scanners operating in the microwave range of the electromagnetic spectrum have very much poorer spatial resolution (from 27 km to 150 km for the SMMR (Scanning Multi-channel Microwave Radiometer) on SEASAT or on NIMBUS-7, see Section 2.5). Better spatial resolution can be achieved with an active microwave system but, as mentioned in Section 2.5, a conventional (or real aperture) radar of the size required cannot be carried on a satellite. Synthetic aperture radar (SAR) provides a solution to the size constraints.

The reconstruction of an image from synthetic aperture radar data is not trivial; it is very expensive in terms of computer time and the theory involved in the development of the algorithms that have to be programmed is complex. This means that the use of synthetic aperture radar involves a sophisticated application of radar system design and signal-processing techniques. Thus a SAR for remote sensing work consists of an end to end system that contains a conventional radar transmitter, an antenna and a receiver together with a processor capable of making an image out of an uncorrelated Doppler phase history. A simplified version of such a system is shown in Figure 7-14. As with any other remote sensing system, the actual design used will depend on the user requirements and on the extent to which it is possible to meet these requirements with the available technology. The system illustrated is based on the earliest implementation technique used to produce images from synthetic aperture radar, that of optical processing. While optical processing has some advantages, it has largely been replaced by electronic (digital) processing techniques.

In considering the Doppler effect in this chapter special attention is given to the case of a moving radar system. The usual expression for Doppler frequency shift is

$$\Delta f = \pm \frac{vf}{c} = \pm \frac{v}{\lambda} \qquad (7\text{-}8)$$

The velocity in this expression is the radial component of the velocity which, in this case, is the velocity of the platform (aircraft or satellite) that is carrying the radar. The positive sign corresponds to the case of approach of source and observer, and the negative sign corresponds to the case of increasing separation between source and observer. For radar there is a two-way transit of the radiowaves between the transmitter and receiver giving a shift of

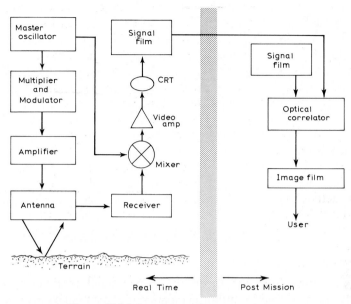

Figure 7-14 Imaging radar operations

$$\Delta f = \pm \frac{2v}{\lambda} \qquad (7\text{-}9)$$

For SAR then, the surfaces of iso-Doppler shift will be cones with their axes along the line of flight of the SAR antenna and with their vertices at the current position of the antenna, see Figure 7-15, while the corresponding iso-Doppler contours on the ground are shown in Figure 7-16.

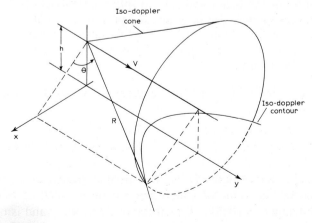

Figure 7-15 Iso-Doppler cone

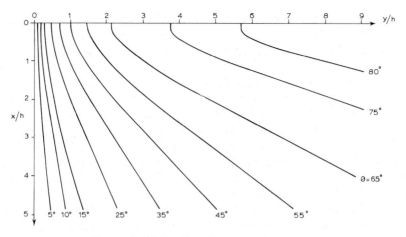

Figure 7-16 Iso-Doppler ground contours

A few points from the theory of conventional (real aperture) radar should be reconsidered for airborne radars rather than radars installed on the ground. For an isotropic antenna radiating power P the energy flux density at a distance R is

$$\frac{P}{4\pi R^2} \tag{7-10}$$

If the power is concentrated into a solid angle Ω instead of being spread out isotropically, the flux will be

$$\frac{P}{\Omega R^2} \tag{7-11}$$

in the direction of the beam and zero in other directions. The half-power beamwidth of an aperture can be expressed as

$$\theta = \frac{1}{\eta} \tag{7-12}$$

where η is the size of the aperture expressed in wavelengths, θ can be expressed as

$$\theta = K\frac{\lambda}{D} \tag{7-13}$$

where λ is the wavelength, D is the aperture dimensions and K is a numerical factor, the value of which depends on the characteristics of the particular antenna in question. K is of the order of unity and is often taken to be equal to one for convenience. For an angular resolution θ the corresponding linear resolution at range R will be given by $R\theta$. If the same antenna is used for both transmission and reception the angular resolution is reduced to θ and the linear resolution becomes $R\lambda/2D$. For a radar system mounted on a moving vehicle this is the along-track resolution. From this formula we see that for a conventional real aperture radar the resolution is better the closer the target is to the radar. Therefore, a long antenna and a short wavelength are required for good resolution.

Now consider the question of the resolution in a direction perpendicular to the direction of motion of the moving radar system. A high resolution radar on an aircraft is mounted so as to be side-looking, rather than looking vertically downward. The acronym SLAR (Side-Looking Airborne Radar) follows from this. The radiation pattern is illustrated in Figure 7-17, i.e., a narrow beam directed at right angles to the direction of flight of the aircraft. A pulse of radiation is transmitted and an image of a narrow strip of the surface of the Earth can be generated from the returns (see Figure 7-18). By the time the next pulse is transmitted and received the aircraft has moved forward a little and another strip of the Earth's surface is imaged. A complete image of the swath AB is built up by the addition of the images of successive strips. Each strip is somewhat analogous to a scan line produced by an optical or infrared scanner. Suppose that the radar transmits a pulse of length $L(=c\tau)$ where τ is the duration of the pulse; then if the system is to be able to distinguish between two objects the reflected pulses must arrive sequentially and not overlap. The objects must therefore be separated by a distance along the ground that is greater than $L/(2 \cos \psi)$ where ψ is the angle between the direction of travel of the pulse and the horizontal. The resolution along the ground in the direction at right angles to the line of flight of the platform, or the range resolution as it is called, is thus

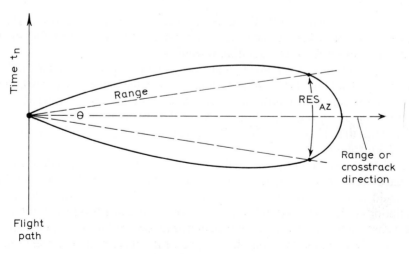

Figure 7-17 Angular resolution

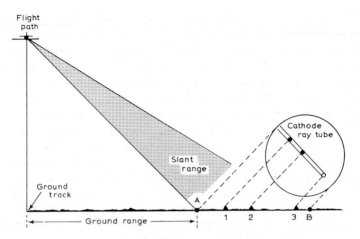

Figure 7-18 Pulse ranging

$c/(2\beta \cos \psi)$ where β, the pulse bandwidth, is equal to $1/\tau$.

It is possible to identify limits on the pulse repetition frequency (PRF) that can be used in a synthetic aperture radar. The Doppler history of a scatterer as the beam passes over it is not continuous but is sampled at the pulse repetition frequency (PRF). The sampling must be at a frequency that is at least twice the highest Doppler frequency in the echo and this sets a lower limit for the PRF. An upper limit is set by the need to sample the swath unambiguously in the range direction, in other words the echoes must not overlap. The PRF limits prove to be

$$\frac{2v}{D} \leq \text{PRF} \leq \frac{c}{2W \cos \psi} \tag{7-14}$$

where W is the swath width along the ground in the range direction. These are very real limits for a satellite system and effectively limit the swath width achievable at a given azimuth resolution.

It is important to realize that a synthetic aperture radar and a conventional real aperture radar system achieve the same range resolution and that the reason for utilizing the technique of aperture synthesis is to improve the along-track resolution. As well as being called the along-track resolution, this is also called the angular cross-range resolution or the azimuth resolution.

It should also be noticed that the range resolution is independent of the distance between the ground and the vehicle carrying the radar. In addition, it should also be noted that we use the term "range resolution" to mean resolution on the ground and at right angles to the direction of flight; it is not a distance along the direction of propagation of the pulses. It can be seen that to increase the range resolution, for a given angle ψ, the pulse duration τ has to be made as short as possible. There is, however, a problem in that it is also necessary to transmit enough power to give rise to a reflected pulse that, on return to the antenna, will be large enough to be detected by the instrument. In order to transmit a given power while shortening the duration of the pulse, the amplitude of the signal must be increased, but it is difficult to design and build equipment to transmit very short pulses of very high energy. A method which is very widely adopted to cope with this problem involves using a "chirp" instead of a pulse of a pure single frequency. A chirp consists of a long pulse with a varying frequency. When the reflected signal is received by the antenna it is fed through a "dechirp network" which produces different delays for the different frequency components of the chirp. This can be thought of as compressing the long chirp pulse into a much shorter pulse of a correspondingly higher amplitude and therefore increasing the range resolution; alternatively it can be thought of as dealing with the problem of overlapping reflected pulses by using the differences in the frequencies to distinguish between them. The idea of using a pulse with a varying frequency in a detection and ranging system is not original to radar systems. It is actually used by some species of bats in an ultrasonic version of a detection and ranging system (see, for example, Cracknell, 1980).

From the expression $R\lambda/2D$ given above for the along-track range it can be seen that to obtain good along-track resolution for a real aperture radar one needs a long antenna, a short wavelength and a close range. There are limits to the lengths of the antennae which can reasonably be carried and stabilized on a satellite or on an aircraft flying at high altitude; moreover, the use of shorter wavelengths will involve greater attenuation of the radiation by clouds and by the atmosphere and thereby reduce the all-weather capability of the radar. While conventional radars are primarily used in short-range operations at low level, synthetic aperture radars were developed to overcome these difficulties and are now used both in aircraft and in satellites.

A SAR has an antenna which is travelling in a direction parallel to its length. The antenna is, in general, moving continuously but the signals transmitted and received back are pulsed. The pulses are transmitted at regular intervals along the flight path; when these individual signals are stored and then added, an antenna of long effective length is synthesized in space. Of course this synthetic antenna is many times longer than the actual antenna and, therefore, gives a much narrower beam and much better resolution. There is an important difference which should be noted between a real aperture antenna and the synthetic antenna. For the real aperture antenna, only a single pulse at a time is transmitted, received and displayed as a line on the image. For the synthetic antenna, each target will produce a large number of return pulses. This set of returns from each target must be stored and then combined in an appropriate manner so that the synthetic antenna can simulate a physical antenna of the same length. The along-track resolution is then determined from theory which is very similar to the theory of the resolution of an ordinary diffraction grating. In place of the spacing between lines on the grating, d represents the distance between successive positions of the transmitting antenna when pulses are transmitted. Without considering the details of the theory we simply quote the result, namely that the along-track resolution, or azimuth resolution, of a SAR is just equal to half the (real) length of the antenna used. This may, at first sight seem rather surprising. However, the distance from the platform to the surface of the Earth is not entirely irrelevant; it must be remembered that it is necessary to transmit enough power to be able to receive the reflected signal.

If a radar is flown on an aircraft one can, for practical purposes, ignore the curvature of the Earth, the rotation of the Earth and the fact that the wavefront of the radar system is a spherical wave and not a plane wave. On the other hand, these factors must be taken into account in a radar system that is flown on a satellite. The range from the radar to an individual scattering point in the target area on the ground will change as the beam passes over the scattering point. This is known as "range walk". There are two components to this effect, one is a quadratic term resulting from the curvature of the Earth and the other is a linear term resulting from the rotation of the Earth. Each point must be tracked through the aperture synthesis to remove this effect, while the actual behaviour for a particular point will depend on the latitude and on the range. In order to compensate for the curvature of the reflected wave front it is necessary to add a range-dependent quadratic phase shift along the synthetic aperture. This is equivalent to focusing the radar at each range gate and if this is not carried out the ground along-track resolution will be degraded to $K\sqrt{\lambda R}$ or, approximately, $\sqrt{\lambda R}$.

One very obvious feature of any synthetic aperture radar image (see Figure 7-19) is a characteristic "grainy" or speckled appearance. The feature is common to all coherent imaging systems and it arises as a result of scattering from a rough surface, i.e. from a surface on which the irregularities are large with respect to the wavelength. This speckle can provide a serious obstacle in the interpretation of synthetic aperture radar images. A technique known as "multi-looking" is commonly used to reduce the speckle. To obtain the best along-track resolution the full Doppler bandwidth of the echoes must be utilized. However, it is possible to use only a fraction of the available bandwidth and produce an image over what is effectively a sub-aperture of the total possible synthetic aperture; this will have a poorer along-track resolution. By using Doppler bands centred on different Doppler frequencies, so that there is no overlap, one can generate a number of independent images for a given scene. Because these different images, or "looks" as they are commonly called, are independent their speckle patterns will also be statistically independent. The speckle can then be very much reduced by incoherently averaging over the different "looks" to produce a multi-look image; however, this is achieved at the expense of

Figure 7-19 SEASAT SAR data of the Tay Estuary, Scotland, from orbit 762 of 19 August 1978 processed digitally (RAE Farnborough)

azimuthal (along-track) resolution. Typically three or four looks are used in producing a multi-look image.

Having considered the question of spatial resolution, a little consideration will be given to the question of image formation using a synthetic aperture radar. It is not appropriate in this book to consider the details of the theory involved in the construction of an image from raw synthetic aperture radar data (for details see, for example, McCord, 1962; Cutrona *et al.*, 1966; Lodge, 1981). Having established the theory, the processing itself can be carried out using either optical or digital techniques. Optical image-processing techniques used for SAR data have the advantage that they produce images very quickly but have the disadvantage that the images are of poorer quality than those produced using digital techniques, (see Figures 7-20 and 7-21).

It is also not within of the scope of this book to enter into extensive discussions of the problems involved in the interpretation of SAR images. In a radar image the intensity at each point represents the radar backscattering coefficient for a point, or small area, on the ground. In a

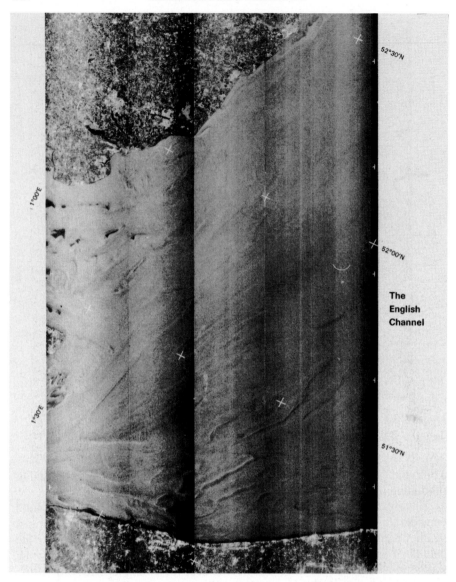

Figure 7-20 Optically processed SEASAT SAR image of the English Channel from orbit 762 of 19 August 1978

photograph or scanner image in the visible or near-infrared region of the electromagnetic spectrum the intensity at a point in the image represents the reflectivity, at the appropriate wavelength, of the corresponding small area on the ground. In a thermal-infrared or passive microwave image the intensity is related to the temperature and the emissivity of the corresponding area on the ground. It is prudent to remember that it should

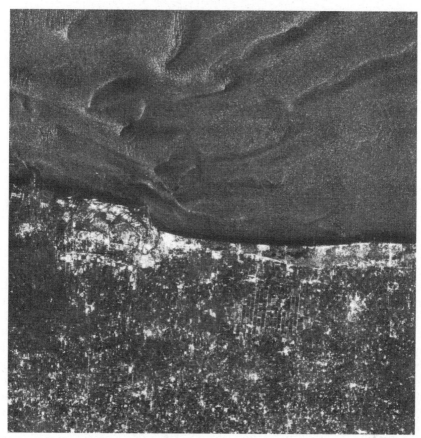

Figure 7-21 Digitally processed data for part of the scene shown in Figure 7-20 (RAE Farnborough)

not be assumed that there is necessarily any simple correlation between images of a given piece of the surface of the Earth produced by such very different physical processes, even if data used to produce the images were generated simultaneously.

8 Atmospheric corrections to passive satellite remote sensing data

8.1 Introduction

Distinction should be made between two types of situation in which remote sensing data are used. In the first type a complete experiment is designed and carried out by a team who are also responsible for all the analysis and interpretation of the data obtained. Such experiments are usually intended either to gather geophysical data or to demonstrate the feasibility of an environmental applications project involving remote sensing techniques. The second type of situation is one in which remotely sensed data are acquired on a speculative basis by the operator of an aircraft or satellite and then distributed to potential users at their request. In this second situation it is necessary to draw the attention of the users to the fact that atmospheric corrections may be rather important if they propose to use the data for environmental scientific or engineering work.

The useful information about the target area of the land, sea or clouds is contained in the physical properties of the radiation leaving that target area whereas what is measured by a remote sensing instrument are the properties of the radiation that arrives at the instrument. This radiation has travelled some distance through the atmosphere and accordingly has suffered both attenuation and augmentation in the course of that journey. The problem that faces the user of remote sensing data is that of being able to regenerate the details of the properties of the radiation that left the target area from the data generated by the remote sensing instrument. An attempt to set up the radiative transfer equation to describe all the various processes that corrupt the signal that leaves the target area on the land, sea or cloud from first principles is a nice exercise in theoretical atmospheric physics and, of course, is a necessary starting point for any soundly based attempt to apply atmospheric corrections to satellite data. However, in a real situation the problem soon arises that suitable values of various atmospheric parameters have to be inserted into the radiative transfer equation in order to arrive at a solution.

8.2 *Radiative transfer theory*

Making quantitative calculations of the difference between the aircraft-received radiance, which is recorded by a set of remote sensing instruments, and the Earth-leaving radiance, which one is trying to measure, is problematical. Expressed in terms of satellites rather than aircraft, one has to try to make quantitative calculations of the difference between the satellite-received radiance and the Earth-leaving radiance. An attempt to solve this problem involves the use of what is commonly known as radiative transfer theory. In essence, this consists of studying radiation travelling in a certain direction, specified by the angle ϕ between that direction and the vertical axis z, and setting up a differential equation for a small horizontal element of the transmitting medium (the atmosphere)with thickness dz. It is necessary to consider

- The radiation entering the element dz from below;

- The attenuation suffered by that radiation within the element dz;

- The additional radiation that is either generated within the element dz or scattered into the direction ϕ within the element dz;

and thence to determine an expression for the intensity of the radiation leaving the element dz in the direction ϕ.

The resulting differential equation is called the radiative transfer equation. Although it is not particularly difficult to formulate, this general form of this equation is not commonly used. In practice, the details of the formulation will be simplified to include only the important effects. The equation will therefore be different for different wavelengths of electromagnetic radiation because of the different relative importance of different physical processes at different wavelengths. Suitable versions of the radiative transfer equation for optical and near-infrared wavelengths, for thermal-infrared wavelengths and for (passive) microwave radiation will be presented in the sections that follow.

If the values of the various atmospheric parameters that appear in the radiative transfer equation are known, this differential equation can be solved to determine the relation between the aircraft-received radiance, or satellite-received radiance, and the Earth-leaving radiance. However, the greatest difficulty in making atmospheric corrections to remotely sensed data lies in the fact that it is usually impossible to obtain accurate values for the various atmospheric parameters that appear in the radiative transfer equation. The atmosphere is a highly dynamic physical system and the various atmospheric parameters will, in general, be functions of the three space variables, x, y, and z, and of the time variable, t. Because of the paucity of the data it is common to assume a horizontally stratified atmosphere, in other words the atmospheric parameters are assumed to be functions of the height z but not the x and y coordinates in a horizontal plane. The situation may be simplified further by assuming that the atmospheric parameters are given by some model atmosphere based only on the geographical location and the time of the year. However, this is not a very realistic approach because the actual atmospheric conditions will differ quite considerably from such a model. It is clearly much better to try and use values of the atmospheric parameters that apply at the time that the remotely sensed data are collected. This can be done

- By using simultaneous, or nearly simultaneous, data from sounding instruments, either radiosondes or satellite-flown sounders;

- By using a multi-channel (i.e. a multi-spectral) approach and, effectively, using a large number of channels of data to determine the atmospheric parameters;

- By using a multi-look approach in which a given element of the surface of the Earth is viewed in rapid succession from a number of different directions, i.e. through different atmospheric paths, so that the atmospheric parameters can either be determined or eliminated from the calculation of the Earth-leaving radiance.

Some examples of these different approaches will be presented in the sections that follow. It is, however, important to realize that there is a fundamental difficulty, namely that the problem of solving the radiative transfer equation in the situations described is an example of an unconstrained inversion problem. That is, there are many unknowns (the atmospheric parameters for a given atmospheric path) and a very small number of measurements (the intensities received in the various spectral bands for the given instantaneous field of view). The solution will, inevitably, take some information from the mathematical and physical assumptions that have been built into the method of solution adopted.

A general formalism for atmospheric absorption and transmission is required. Consider a beam of radiation with wavelength λ and wavenumber $\kappa (= 2\pi/\lambda)$ travelling at a direction θ to the normal to the Earth's surface. After the radiation has travelled a distance l, the radiance flux (radiance) of the wavelength λ, $\phi_\lambda(l)$, is related to its initial value $\phi_\lambda(0)$ by

$$\phi_\lambda(l) = \phi_\lambda(0) \exp\left\{ -\sec\theta \int_0^l K_\lambda(z)\, dz \right\} \qquad (8\text{-}1)$$

where $z = l \cos\theta$ and $K_\lambda(z)$ is the attenuation coefficient. Notice that the attenuation coefficient is a function of height as well as of wavelength. These quantities can be expressed in terms of κ instead of λ giving

$$\phi_\kappa(l) = \phi_\kappa(0) \exp\left\{-\sec\theta \int_0^l K_\kappa(z)\,dz\right\} \tag{8-2}$$

The dimensionless quantity

$$\int_0^z K_\kappa(z)\,dz$$

is called the optical thickness and is commonly denoted by $\tau_\kappa(z)$ and the quantity

$$\exp\left(-\int_0^z K_\kappa(z)\,dz\right)$$

is called the beam transmittance and is commonly denoted by $T_\kappa(z)$.

8.3 *Physical processes involved in atmospheric correction*

In atmospheric correction processes, the first distinction to be made is whether the radiation leaving the surface of the land, sea or clouds is radiation emitted by that surface or whether it is reflected solar radiation. The relative proportions of reflected and emitted radiation will vary according to the wavelength of the radiation, the time and the place of observation. It has already been noted in Section 2.2 that at optical and very near-infrared wavelengths, the emitted radiation is negligible compared with the reflected radiation, while at thermal-infrared and microwave wavelengths it is the emitted radiation which is important with the reflected radiation being of negligible intensity. Within the limitations of the accuracy of these estimates it may be seen that at a wavelength of 3.5 μm, which is actually the wavelength of one of the bands of the AVHRR, emitted and reflected radiation are both important. The problem is to relate data usually consisting of, or derived from, the output from a passive scanning instrument, to the properties of the land, sea or clouds under investigation.

The approach adopted to the question of the contribution of the intervening atmosphere to remotely-sensed data is governed both by the characteristics of the remote sensing system in use and by the nature of the environmental problem to which the data are to be applied. In work that has been done so far in land-based applications of remote sensing it has been relatively rare to take much notice of atmospheric effects while in meteorological applications it is the atmosphere itself that is the object of investigation. A great deal of meteorological information can be extracted from remote sensing data without needing to take any account of details of the corruption of the signal from the target by intervening layers of the atmosphere. Thus images from systems such as the TIROS-N series of polar-orbiting satellites or from the geostationary satellites such as METEOSAT, GOES-E and GOES-W can be used to give synoptic views of whole weather systems and their developments in a manner that was completely impossible previously. If experiments are being conducted to study the physical properties and motions of a given layer of the atmosphere it may be necessary to make allowance for contributions to a remotely sensed signal from other atmospheric layers. The area of work in which, so far, atmospheric effects have been of greatest concern to the users of remote sensing data have been those in which water bodies, lakes, lochs, rivers and the oceans have been studied in order to determine the physical or biological parameters of the water.

In most cases the user of remote sensing data is interested in knowing how important the various atmospheric effects are on the quality of image data or on the magnitudes of derived physical or biological parameters; the user is not usually interested in the magnitudes of the corrections to the radiance values *per se*. However, to assess the relative importance of the various atmospheric effects it is necessary to devote some attention to

- The physical processes occurring in the atmosphere;

- The magnitudes of the effects of these processes on the radiance reaching the satellite; and

- The consequences of these effects on images or on derived physical or biological parameters.

There are several different approaches that one can take to the question of applying atmospheric corrections to satellite remote sensing data for the extraction of geophysical parameters. We note the following options

1. Ignore atmospheric effects completely;

2. Calibration with *in situ* measurements of geophysical parameters;

3. The use of a model atmosphere with parameters determined from historic data;

4. The use of a model atmosphere with parameters determined from simultaneous meteorological data;

5. The elimination of, or compensation for, atmospheric effects on a pixel-by-pixel basis.

The selection of the appropriate option will be governed by considerations both of the sensor that is being used to gather the data and of the problem to which the data are being applied.

1. One can ignore the atmospheric effects completely. This is not quite so frivolous or irresponsible as it might seem at first sight. In practice we find that it is the case that there are some

applications for which this is often a perfectly acceptable approach.

2. One can calibrate the data with the results of some simultaneous *in situ* measurements of the geophysical parameter that one is trying to map from the satellite data. These *in situ* measurements may be obtained for a training area or at a number of isolated points in the scene. It is important that the measurements are actually made simultaneously with the gathering of the data by the satellite.

Many examples of the use of this approach can be found for data both from the visible channels and from the infrared channels of aircraft-flown and satellite-flown scanners and some relevant references are cited in Sections 8.4.3 and 8.5.1. The method of calibration with simultaneous *in site* data is capable of yielding quite accurate results. It is quite successful in practice although, of course, the value of remote sensing techniques can be considerably enhanced if the need for simultaneous *in situ* calibration data can be eliminated. Having used simultaneous *in situ* data to calibrate scanner data for a given geographical area on one day then one might try to use the same calibration "constants" for the same geographical area on another day. However, the evidence available at present seems to suggest that this is not the case. An example to illustrate this is provided by Singh and Warren (1983) for AVHRR data and sea-surface temperatures. A similar situation is found in the case of CZCS data. Suppose that we take a given form of algorithm for extracting chlorophyll concentrations from CZCS data. Then we find that calibration "constants" for chlorophyll concentrations found using data for one day, if applied to CZCS data for the same geographical area but from another day, may predict values of chlorophyll concentrations that are in error by almost an order of magnitude (Sathyendranath and Morel, 1983).

In addition to the problems associated with variations in the atmospheric conditions from day to day there is also the quite serious problem that there are likely to be significant variations in the atmospheric conditions even within a given scene at any one time. To allow for the variations that may exist between atmospheric conditions in different parts of a given scene it would be necessary to have available *in situ* calibration data for a much finer network of closely packed points than would be at all feasible. While it is, of course, necessary to have some *in situ* data available for initial validation checks and for occasional monitoring thereafter on results derived from satellite data, to use a large network of *in situ* calibration data largely negates the value of using remote sensing data anyway, since one important objective of using remote sensing data was largely to eliminate costly field work.

From what we have said it is clear that a great deal more experience needs to be acquired before one is going to be able to determine values of geophysical parameters from satellite data for which results of simultaneous *in situ* measurements are not available. It is this sort of difficulty which has led to the adoption of methods that involve trying to eliminate atmospheric effects rather than trying to calculate the atmospheric corrections.

3. One can use a model atmosphere, with the details and parameters of the model adjusted according to the geographical location and the time of the year. This is more likely to be successful if one is dealing with an instrument with low spatial resolution that is gathering data over wide areas for an application that involves taking a global view of the surface of the Earth. In this situation the local spatial irregularities and rapid temporal variations in the atmosphere are likely to cancel out and fairly reliable results may be obtained. This approach is also likely to be relatively successful for situations in which the magnitude of the atmospheric correction is relatively small compared with the signal from the target area that is being observed. All these conditions are satisfied for passive microwave radiometry and so this approach is moderately successful for SMMR data (see, for example, Thomas, 1981, and references therein).

4. One can also use a model atmosphere but make use of such simultaneous meteorological data as may actually be available instead of using only assumed values based on geographical location and time of year. This simultaneous meteorological data may be obtained from one of several possible sources. The satellite may, like the TIROS-N series of satellites, carry other instruments, in addition to the scanner, which are used for carrying out meteorological sounding through the atmosphere below the satellite, see Section 8.4.

5. One can attempt to eliminate atmospheric effects. This can be done in various ways. For instance, one can try to eliminate the effect of the atmosphere by a multi-look approach in which a given target area on the surface of the sea is viewed

from two different directions. This has been tried using data from two different satellites (e.g. Chedin *et al.*, 1982; Holyer, 1984) though this has the disadvantage that the two paths are very different from one another and the cancellation of atmospheric effects between the two paths is then not likely to be very accurate. Instead of looking from two different satellites one can use an instrument such as the Along Track Scanning Radiometer (ATSR/M) which it is planned to fly on ERS-1; this instrument by looking both forward and vertically downwards will view a given target area on the sea surface twice in rapid succession through two different atmospheric paths and the intention will be to eliminate the atmospheric effects between these two paths. Alternatively, instead of the multi-look approach, one can attempt to eliminate atmospheric effects by exploiting a number of different spectral channels to try to cancel out the atmospheric effects between these channels. In principle, if one has enough spectral channels, this method should be capable of giving very good results because it is using identical atmospheric paths for the channels that are being compared. For this purpose the version of the AVHRR used on the latest satellites in the TIROS-N series has two channels close together in the wavelength range of 10 – 12 μm. The correction of the data from the visible bands of the CZCS also involves using several channels essentially to eliminate atmospheric effects.

Most of the options presented will be considered in relation to microwave, thermal-infrared and visible-band data in the sections that follow. The cases of emitted radiation and reflected solar radiation are considered separately, with consideration also being given to atmospheric transmittance.

8.3.1 Emitted radiation

In Table 2-1 it can be seen that at long wavelengths, i.e. for microwaves and for thermal-infrared radiation, it is the emitted radiation, not the reflected solar radiation, that is important. There are several contributions to the radiation received at the instrument, see Figure 8-1; these contributions, identified as T_1, T_2, T_3 and T_4, are described in the following subsections 8.3.1.1–8.3.1.4. Each can be considered as a radiance $L(\kappa)$, where κ is the wavenumber, or as corresponding to an equivalent black body temperature.

8.3.1.1 Surface radiance: $L_1(\kappa), T_1$

The surface radiance is the radiation which is generated thermally at the

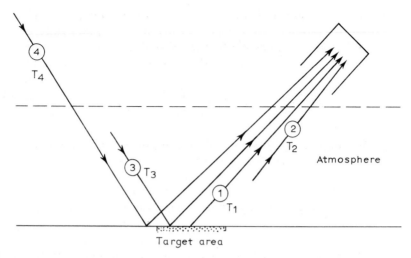

Figure 8-1 Contributions to satellite-received radiance for emitted radiation

Earth's surface and which undergoes attenuation as it passes through the atmosphere before reaching the scanner; this radiance can be written as $\epsilon B(\kappa,T_s)$ where ϵ is the emissivity, $B(\kappa,T_s)$ is the Planck distribution function and T_s is the temperaturre of the surface. In general, the emissivity ϵ is a function of wavenumber and temperature. For example, the emissivity of gases varies very rapidly with wavenumber in the neighbourhood of the absorption (emission) lines. For sea water ϵ may be treated as constant with respect to κ and T_s. If the presence of any material which is not part of sea water is ignored, e.g. oil pollution, industrial waste, etc., then ϵ may be regarded as a constant. Let p_0 be the atmospheric pressure at the sea surface. By definition, the pressure on the top of the atmosphere is zero. Thus, the radiance reaching the detector from the view angle θ is

$$L_1(\kappa) = \epsilon B(\kappa,T_s)\tau(\kappa,\theta;p_0,0) \tag{8-3}$$

where $\tau(\kappa,\theta;p,p_1)$ is the atmospheric transmittance for wavenumber κ and direction θ between heights in the atmosphere where the pressures are p and p_1.

8.3.1.2 *Upwelling atmospheric radiance: $L_2(\kappa),T_2$*

The atmosphere emits radiation at all altitudes. As this emitted radiation travels upwards to the scanner it undergoes attenuation in the overlying atmosphere. It is possible to show (see, e.g., Singh and Warren, 1983) that the radiance emitted by a horizontal slab of the atmosphere lying between heights z and $z+\delta z$, where the pressure is p and $p+\delta p$ respectively, and arriving in a direction θ at a height z_1 where the pressure is p_1, is given by

$$dL_2(\kappa) = B(\kappa,T(p))\ d\tau(\kappa,\theta;p,p_1) \tag{8-4}$$

or

$$dL_2(\kappa) = B(\kappa,T(p))\ \frac{d\tau}{dp}\ (\kappa,\theta;p,p_1)\ dp \tag{8-5}$$

The upwelling emitted radiation received at the satellite can thus be written as

$$L_2(\kappa) = \int_{p_0}^{p} B(\kappa,T(p))\ \frac{d\tau}{dp}\ (\kappa,\theta;p,0)\ dp \tag{8-6}$$

where p_0 is the atmospheric pressure at the sea surface and $T(p)$ is the temperature at the height at which the pressure is p. This expression is based on the assumption of local thermodynamic equilibrium and the use of Kirchhoff's law to relate the emissivity to the absorption coefficient.

8.3.1.3 Downwelling atmospheric radiance: $L_3(\kappa),T_3$

In this case allowance is made for atmospheric emission downwards to the Earth's surface where the radiation undergoes reflection upwards to the scanner. Attenuation is undergone as the radiation passes through the atmosphere. The total downwelling radiation from the top of the atmosphere, where $p=0$, to the sea surface, where the pressure is p_0, is given by

$$\int_{0}^{p_0} B(\kappa,T(p))\{d\tau(\kappa,\theta;p,p_0)/dp\}\ dp \tag{8-7}$$

A fraction $(1-\epsilon)$ of this is reflected at the sea surface and a fraction $\tau(\kappa,\theta;p_0,0)$ of this passes through the atmosphere so that the radiance reaching the satellite is given by

$$L_3(\kappa) = (1 - \epsilon)\tau(\kappa,\theta;p_0,0) \int_{0}^{p_0} B(\kappa,\tau(p))\ \frac{d\tau}{dp}\ (\kappa,\theta;p,p_0)\ dp \tag{8-8}$$

8.3.1.4 Space component: $L_4(\kappa),T_4$

Space has a background brightness temperature of about 3 K. The space component passes down through the atmosphere, is reflected at the surface, and passes up through the atmosphere again to reach the scanner.

8.3.1.5 *Total radiance: $L^*(\kappa)$, T_4*

The total radiance $L^*(\kappa)$ received at the satellite can be written as

$$L^*(\kappa) = L_1(\kappa) + L_2(\kappa) + L_3(\kappa) + L_4(\kappa) \qquad (8-9)$$

Alternatively the same relation can be expressed in terms of the brightness temperature T_b and the equivalent temperatures for each of the contributions already mentioned, i.e.

$$T_b = T_1 + T_2 + T_3 + T_4 \qquad (8-10)$$

8.3.1.6 *Calculation of sea-surface temperature*

Sea-surface temperatures are studied quite extensively using both infrared and microwave passive instruments. In both cases the problem is to estimate or eliminate T_2, T_3 and T_4 so that T_1 can be determined from the measured valued of T_b. There is a further complication in the case of microwave radiation as, for certain parts of the Earth's surface at least, a significant contribution also arises from microwaves generated artificially for telecommunications purposes. It is simplest, from the point of view of the above scheme, to include this contribution in T_2.

Apart from information about the equivalent black body temperature of the surface of the land, sea or cloud, the brightness temperature measured by the sensor contains information on a number of atmospheric parameters such as water vapour content, cloud liquid water content and rainfall rate. By using multi-channel data it may be possible to eliminate T_2, T_3 and T_4 and hence to calculate T_1 from T_b.

8.3.2 Reflected radiation

The reflected radiation case concerns radiation that originates from the Sun and eventually reaches a remote sensing instrument on an aircraft or spacecraft, the energy of the radiation that arrives at the instrument being measured by the sensor. Hopefully the bulk of this radiation will come from the instantaneous field of view, IFOV, on the target area of land, sea or cloud that is the observed object of the remote sensing activity. However, in addition to radiation that has travelled directly over the path Sun → IFOV → sensor and which may contain information about the area that is seen in the IFOV, there will also be some radiation which reaches the sensor by other routes. This radiation will clearly not contain information about the IFOV. Accordingly various paths between the Sun and the sensor are considered for reflected radiation reaching the sensor, (see Figures 8-2 and 8-3):

1. $L_1(\kappa)$: radiation that follows a direct path from the Sun to the target area and thence to the sensor;

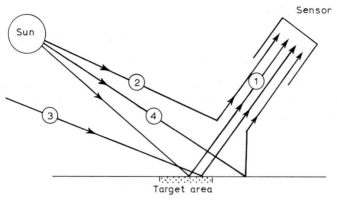

Figure 8-2 Contributions to satellite-received radiance for reflected solar radiation; 1, 2, 3 and 4 denote $L_1(\kappa)$, $L_2(\kappa)$, $L_3(\kappa)$ and $L_4(\kappa)$ respectively

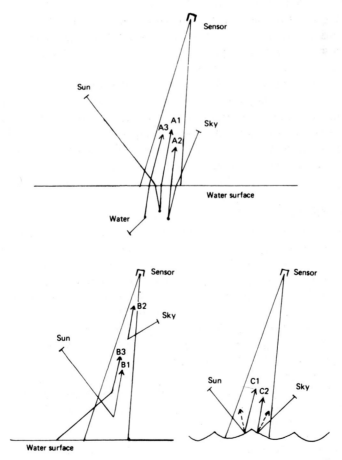

Figure 8-3 Components of the sensor signal in remote sensing of water (Sturm, 1981)

2. $L_2(\kappa)$: radiation from the Sun that is scattered into the field of view of the sensor, either by single or multiple scattering in the atmosphere, without the radiation ever reaching the target area at all;

3. $L_3(\kappa)$: radiation which does not come directly from the Sun but which has first undergone some scattering event before reaching the target area; this radiation then passes to the sensor directly;

4. $L_4(\kappa)$: radiation which has been reflected by other target areas of the land, sea or clouds and is then scattered by the atmosphere into the field of view of the sensor.

These four processes may be regarded, to some extent, as analogues, for reflected radiation, of the four processes outlined in Section 8.3.1 for emitted radiation.

It is $L_1(\kappa)$ that contains the useful information. $L_2(\kappa)$ and $L_4(\kappa)$ do not contain useful information about the target area. While, in principle, $L_3(\kappa)$ does contain some information about the target area it may be misleading information if the radiation is mistakenly regarded as having travelled directly from the Sun to the target area.

It cannot be assumed that the spectral distribution of the radiation reaching the outer regions of the Earth's atmosphere, or its intensity integrated over all wavelengths, is constant. The extraterrestrial solar spectral irradiance (as it is called) and its integral over wavelength, which is called the solar constant, have been studied experimentally over the last 50 years or more. The technique which is used is due originally to Langley and involves the extrapolation of ground-based irradiance measurements to outside the Earth's atmosphere. A review of such measurements, together with recommendations of standard values was given by Labs and Neckel (1967, 1968, 1970). Measurements of the extraterrestrial irradiance have also been made from an aircraft flying at a height of 11.6 km (Thekaekara *et al.*, 1969). While various experimenters acknowledge errors in the region of ±3% following the calibration of their instruments to radiation standards, the sets of results differ from one another by considerably more than this; in some parts of the spectrum they differ by as much as 10%. Some results are shown in Figure 8-4. Some of the discrepancy is explained by the fact that the radiation from the Sun itself varies. Annual fluctuations in the radiance received at the Earth's atmosphere associated with the variation of the distance from the Sun to the Earth can be taken into account mathematically. The eccentricity of the ellipse describing the orbit of the Earth is 0.0167 and the minimum and maximum distances from the Sun to the Earth occur on 3 January and 2 July, respectively. The extraterrestrial solar irradiance for Julian day D is given by the following expression:

$$E_0(D) = \bar{E}_0[1 + 0.167 \cos\{(2n/365)(D - 3)\}]^2 \qquad (8\text{-}11)$$

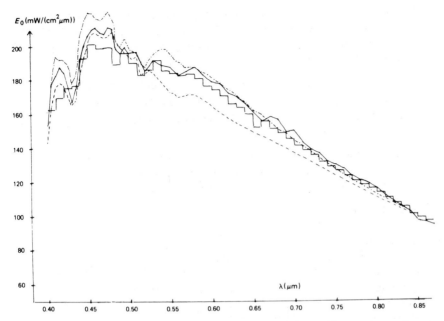

Figure 8-4 Solar extraterrestrial irradiance (averaged over the year) as a function of wavelength (from four different sources, after Sturm, 1981)

8.3.3. Atmospheric transmission

The possible origins of the radiation that finally reaches a remote sensing instrument, and the possible routes that the radiation may take in travelling from its source to the sensor, were considered in Sections 8.3.1 and 8.3.2. It is also necessary to consider the scattering mechanisms involved, both in the atmosphere and at the target area on the surface of the Earth or the clouds. While the reflection or scattering at the target area is relevant to the use of all remotely-sensed data, the details of the interaction of the radiation with the target area are not considered here, where attention is focused on the scattering and absorption of the radiation which occurs during the passage of radiation through the atmosphere.

Three types of scattering are distinguished depending on the relationship between a, the diameter of the scattering particle, and λ, the wavelength of the radiation. If $a \ll \lambda$ Rayleigh scattering is dominant. For Rayleigh scattering the scattering cross section is proportional to $1/\lambda^4$; for visible radiation this applies to scattering by gas molecules. Other cases correspond to scattering by aerosol particles. If $a \approx \lambda$ Mie scattering is dominant. Mie scattering involves water vapour and dust particles. If $a \gg \lambda$ non-selective scattering is dominant. This scattering is independent of wavelength; for the visible range this involves water droplets with radii of the order of $5 - 100$ μm.

The mechanisms involved in scattering or absorption of radiation as it passes through the atmosphere can be conveniently considered as follows. The attenuation coefficient $K_\kappa(z)$ mentioned in Section 8.2 can be separated into two parts

$$K_\kappa(z) = K_\kappa^M(z) + K_\kappa^A(z) \tag{8-12}$$

where $K_\kappa^M(z)$ and $K_\kappa^A(z)$ refer to molecular and aerosol attenuation coefficients. Each of these absorption coefficients can be written as the product of $N^M(z)$ or $N^A(z)$, the number of particles per unit volume at height z, and a quantity σ_κ^M or σ_κ^A, known as the effective cross section, i.e.

$$K_\kappa(z) = N^M(z)\sigma_\kappa^M + N^A(z)\sigma_\kappa^A \tag{8-13}$$

The quantities

$$\tau_\kappa^M(z) = \sigma_\lambda^M \int_0^z N^M(z)\,dz \tag{8-14}$$

and

$$\tau_\kappa^A(z) = \sigma_\lambda^A \int_0^z N^A(z)\,dz \tag{8-15}$$

are called the molecular optical thickness and the aerosol optical thickness, respectively. It is convenient to separate the molecular optical thickness into a sum of two components

$$\tau_\kappa^M(z) = \tau_{\kappa_s}^M(z) + \tau_{\kappa_a}^M(z) \tag{8-16}$$

where $\tau_{\kappa_s}^M(z)$ corresponds to scattering and $\tau_{\kappa_a}^M(z)$ corresponds to absorption. Thus the total optical thickness can be written as

$$\tau_\kappa(z) = \tau_{\kappa_s}^M(z) + \tau_{\kappa_a}^M(z) + \tau_\kappa^A(z) \tag{8-17}$$

These three contributions are considered briefly in turn.

8.3.3.1 Scattering by air molecules

At optical wavelengths this involves Rayleigh scattering. The Rayleigh scattering cross section is given by a well-known formula

$$\sigma_{\lambda_s}^M = \frac{8\pi^3(n^2 - 1)^2}{(3N^2\lambda^4)} \tag{8-18}$$

where n = refractive index, N = number of air molecules per unit volume, and λ = wavelength. This contribution to the scattering of the radiation can be calculated in a relatively straightforward manner. The λ^{-4} behaviour of the Rayleigh scattering (molecular scattering) means this mechanism is very important at short wavelengths but becomes unimportant at long wavelengths. The blue colour of the sky and the red colour of sunrises and unsets is attributable to the difference between this scattering for blue light and red light. This mechanism becomes negligible for near-infrared wavelengths, see Figure 8-5, and is of no importance for microwaves.

8.3.3.2 Absorption by gases

In remote sensing work it is usual to use radiation of wavelengths that are not within the absorption bands of the major constituents of the atmosphere. The gases to be considered are O_2 and N_2, the main constituents of the atmosphere, and carbon dioxide, ozone and water vapour. At optical wavelengths the absorption by O_2, N_2 and CO_2 is negligible. Water vapour has a rather weak absorption band for wavelengths from about 0.7 to 0.74 μm. The only significant contribution to atmospheric absorption by molecules is by ozone. This contribution can be calculated and, although it is small in relation to the Rayleigh and aerosol contribution, it should be included in any calculations of atmospheric corrections to optical scanner data, (see Table 8-1). For scanners operating in the thermal-infrared and microwave regions of the electromagnetic spectrum absorption by gases constitutes the major absorption mechanism. The attenuation experienced by the radiation can be calculated using the absorption spectra of the gases involved, carbon

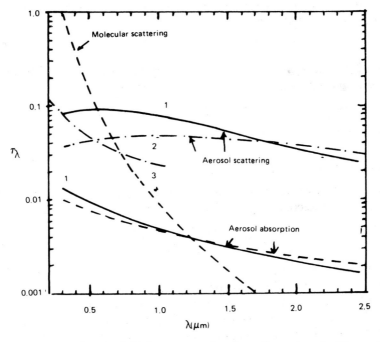

Figure 8-5 Normal optical thickness as a function of wavelength (Sturm, 1981)

Table 8.1 Ozone optical thickness for vertical path through the entire atmosphere (after Sturm, 1981)

Wavelength $\lambda(\mu m)$	Ozone abs. coefficient $\kappa^0_\lambda(cm^{-1})$	Atmosphere type				
		1 $V^0(\infty)=0\cdot23$	2 $V^0(\infty)=0\cdot39$	3 $V^0(\infty)=0\cdot31$	4 $V^0(\infty)=0\cdot34$	5 $V^0(\infty)=0\cdot45$
0·44	0·001	0·0002	0·0004	0·0003	0·0003	0·0005
0·52	0·055	0·0128	0·0213	0·0173	0·0187	0·0245
0·55	0·092	0·0215	0·0356	0·0289	0·0312	0·0409
0·67	0·036	0·0084	0·0139	0·0113	0·0122	0·0160
0·75	0·014	0·0033	0·0054	0·0044	0·0048	0·0062

$V^0(\infty)$ is the visibility range parameter which is related to the optical thickness $\tau^0_\lambda(\infty)$ and the absorption coefficient κ^0_λ for ozone by $\tau^0_\lambda(\infty) = \kappa^0_\lambda V^0(\infty)$

dioxide, ozone and water vapour, (see Figure 8-6). The relative importance of the contributions from these three gases will depend on the wavelength range under consideration. As indicated, it is only the ozone absorption that is significant at visible wavelengths.

8.3.3.3 Scattering by aerosol particles

The aerosol scattering also decreases with increasing wavelength. It is common to write the aerosol optical thickness as

$$\tau_\lambda^A = A\lambda^{-B} \qquad (8\text{-}19)$$

where B is referred to as the Ångström exponent. However, the values of the parameters A and B do vary quite considerably according to the nature of the aerosol particles. Quoted values of B vary from 0.8 to 1.5 or even higher. At optical wavelengths the aerosol scattering is comparable in magnitude with the Rayleigh scattering, see Figure 8-5. In practice, however, it is more difficult to calculate because of the great variability in the nature and concentration of aerosol particles in the atmosphere. Indeed, when dealing with data from optical scanners it is the calculation of the aerosol scattering which is the most troublesome part of the atmospheric correction calculations. While being of some importance in the near infrared, aerosol scattering can be ignored in the thermal infrared for clear air, i.e. in the absence of cloud, haze, fog or smoke, and it can be ignored in the microwave region.

Estimates of corrections to remotely sensed data are based, ultimately, on solving the radiative transfer equation although, as has been indicated in the previous section, accurate solutions are very hard to obtain and one is forced to adopt an appropriate level of approximation.

The importance of understanding the effect of the atmosphere on remote

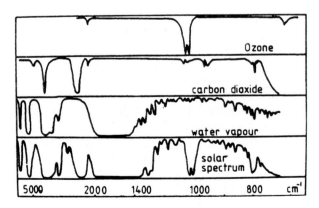

Figure 8-6 The absorption spectra of ozone, carbon dioxide, water vapour and of all atmospheric constituents (Singh and Warren, 1983)

sensing data and of making corrections for atmospheric effects depends
very much on the use that is to be made of the data. There are many
meteorological and land-based applications of remote sensing listed in
Table 1-2 for which in the past there has been no need to carry out any kind
of atmospheric correction. This is either because the information that is
being extracted is purely qualitative or because, though being quantitative,
the remotely-sensed data are calibrated by the use of *in situ* data within a
training area. Nevertheless it is anticipated that in the future some of these
studies will become more exact, particularly as more careful multi-temporal
studies are undertaken of environmental systems that exhibit change. This
is likely to mean that it will become increasingly important to include
atmospheric corrections for some of these applications in the future. In the
case of oceanographic and coastal work the information for extraction
consists of quantitative values of physical or biological parameters of the
water such as the surface temperature and concentrations of suspended
sediment or chlorophyll. While it is interesting to consider the importance
of atmospheric effects in terms of the magnitude of the attenuation relative
to the magnitude of the signal from the target area, these effects should not
be considered in isolation, but should rather be considered in conjunction
with the use to which the data are to be applied. It must also be
remembered that this section is only concerned with passive sensors.

8.4 Passive microwave scanners and thermal-infrared scanners

8.4.1 The radiative transfer equation

For a passive microwave scanner and for a thermal-infrared scanner we
are concerned with emitted radiation and we consider the radiative transfer
equation in the following form

$$\frac{dI_f(\theta,\phi)}{ds} = -\gamma_f I_f(\theta,\phi) + \psi_f(\theta,\phi) \tag{8-20}$$

$I_f(\theta,\phi)$ is the intensity of electromagnetic radiation of frequency f in the
direction (θ,ϕ), s is measured in the direction (θ,ϕ) and γ_f is an extinction
coefficient. The first term on the right-hand side of this equation describes
the attenuation of the radiation both by absorption and by scattering out of
the direction (θ,ϕ). The second term describes the augmentation of the
radiation, both by emission and by scattering of additional radiation into the
direction (θ,ϕ); this term can be written in the form

$$\psi_f(\theta,\phi) = \psi_f^A(\theta,\phi) + \psi_f^S(\theta,\phi) \tag{8-21}$$

where $\psi_f^A(\theta,\phi)$ is the contribution corresponding to the emission and can,
in turn, be written in the form

$$\psi_f^A(\theta,\phi) = \gamma_f^A B_f(T) \tag{8-22}$$

where γ_f^A is an extinction coefficient and $B_f(T)$ is the Planck distribution function for black-body radiation

$$B_f(T) = \frac{2hf^3}{c^2} \frac{1}{\exp(hf/kT) - 1} \tag{8-23}$$

and h = Planck's constant, c = velocity of light in free space, k = Boltzmann's constant and T = absolute temperature. $\psi_f^S(\theta,\phi)$ is the contribution to scattering into the direction (θ,ϕ) and can be written in the form

$$\psi_f^S(\theta,\phi) = \gamma_f^S J_f(\theta,\phi) \tag{8-24}$$

where $J_f(\theta,\phi)$ is a function that depends on the scattering characteristics of the medium. Accordingly, equation 8-20 can be rearranged to give

$$-\frac{1}{\gamma_f} \frac{dI_f}{ds}(\theta,\phi) = I_f(\theta,\phi) - \frac{\gamma_f^A}{\gamma_f} B_f(T) - \frac{\gamma_f^S}{\gamma_f} J_f(\theta,\phi) \tag{8-25}$$

or

$$\frac{dI_f}{d\tau}(\theta,\phi) = I_f(\theta,\phi) - (1 - \omega)B_f(T) - \omega J_f(\theta,\phi) \tag{8-26}$$

where $d\tau = -\gamma_f ds$, τ = optical thickness, $\gamma = \gamma_f^A + \gamma_f^S$ and $\omega = \gamma_f^S/\gamma_f$.

The differential equation is then expressed in terms of optical thickness τ rather than the geometrical path length s.

At microwave frequencies, where $h_f \ll kT$ and the Rayleigh-Jeans approximation can be made, namely that

$$B_f(T) \simeq \frac{2hf^3}{c^2} \frac{1}{(1 + hf)/kT - 1} = \frac{2f^2 kT}{c^2} = \frac{2kT}{\lambda^2} \tag{8-27}$$

then equation 8-27 can be integrated and expressed in terms of equivalent temperatures for black-body radiation

$$T_B(\theta,\phi,0) = T_B(\theta,\phi,\tau)\, e^{-\tau} + \int_0^\tau T_{eff}(\theta,\phi,\tau')\, e^{-\tau'} d\tau' \tag{8-28}$$

where

$$T_{eff}(\theta,\phi,\tau') = [1 - \omega(\tau')] T_m(\tau') + \omega(\tau') T_{sc}(\theta,\phi,\tau') \tag{8-29}$$

and $T_m(\tau')$ is the radiation temperature of the medium and $T_{sc}(\theta,\phi,\tau')$ is a temperature equivalent for the total radiation scattered into the direction (θ,ϕ) from all directions.

There are two slightly different ways in which the problem of solving the radiative transfer equation is likely to be approached. We have introduced it in terms of thinking about it as a means to correct satellite-received radiances, or aircraft-received radiances, to determine the Earth-surface-leaving radiance. In this case one must either have independent data on the physical parameters of the atmosphere or one must assume some values for these parameters. Alternatively, the radiative transfer equation may be used in connection with attempts to determine the atmospheric profile or conditions as a function of height. Atmospheric profiles have been determined for many years by radiosondes which are launched at regular intervals by weather stations. Each radiosonde consists of a balloon, a set of instruments to measure parameters such as pressure, temperature and humidity and a radio transmitter to transmit the data back to the ground. However, since radiosonde stations are relatively sparse, use may also be made of sounding instruments flown on various satellites for determining atmospheric profiles. Perhaps the best-known of these sounding instruments are the TOVS (TIROS Operational Vertical Sounder) series flown on the TIROS-N, NOAA series of weather satellites. These are essentially microwave and infrared multi-spectral scanning or profiling instruments with extremely low spatial resolution. The TOVS system has three separate instruments that are used to determine temperature profiles from the surface to the 50 km level. The High Resolution Infrared Radiation Sounder (HIRS/2) operates in 20 spectral channels, 19 in the infrared and 1 in the visible, at a spatial resolution of 25 km and is mainly used for determining tropospheric temperature and water vapour variations. The four channel Microwave Sounding Unit (MSU) operates at ±54 GHz at which frequency clouds are essentially transparent, although rain causes attenuation. It has a spatial resolution of 110 km and is the major source of information when the sky is overcast. The Stratospheric Sounding Unit (SSU) is a three channel infrared instrument for measuring temperatures in the stratosphere (25 – 50 km) at a spatial resolution of 25 – 45 km. These sounding instruments are described in NOAA reports (Schneider *et al.*, 1981; Werbowetzki, 1981) and, until the split-channel version of the AVHRR was introduced on NOAA-7 (see option 5 outlined in Section 8.3), data from these sounding instruments were used routinely by NOAA in the application of atmospheric corrections to AVHRR data for the production of sea-surface temperature charts.

The data from such sounding instruments are usually analysed by neglecting the scattering into the direction (θ, ϕ) so that equations 8-20 and 8-28 can be simplified by neglecting the scattering, i.e. by setting ω or $\omega(\tau')$ (all τ') equal to zero. Thus, on integrating equation 8-25 from zero (at the surface) to infinity (at the satellite) in this approximation we obtain

$$I_f(\theta, \phi) = B_f(T_s)\tau_f(0, \infty) + \int_0^\infty B_f(T(z)) \frac{d\tau_f(z, \infty)}{dz} \, dz \qquad (8\text{-}30)$$

The quantity $d\tau_f(z, \infty)/dz$ may be written as $K_f(z)$ for convenience. Using equation 8-30 to give the intensity of radiation $I_f(\theta,\phi)df$ that is received in a (narrow) spectral band of width df we have

$$I_f(\theta,\phi) \ df = B_f(T_s)\tau_f(0,\infty) \ df + \int_0^\infty B_f(T(z))K_f(z) \ dzdf \quad (8\text{-}31)$$

Information gathered by a sounding instrument is then used to invert the set of values of $I_f(\theta,\phi)$ df from the various spectral bands of the instrument to determine the temperature profile. This involves making use of a given set of values of $K_f(z)$, which may be regarded as a weighting function. In practice it is usual, however, to transform this weighting function to express it in terms of pressure, p, rather than height, z, and also to express the temperature profile derived from the sounding measurements as a function, $T(p)$, of pressure rather than as a function, $T(z)$, of height.

8.4.2 The Scanning Multi-channel Microwave Radiometer

The SMMR (Scanning Multi-channel Microwave Radiometer), which was flown on SEASAT and on NIMBUS-7 and which was mentioned in Section 2.5 is particularly suited to global oceanographic work because of its large instantaneous field of view and large area of coverage. The data are used to extract estimates of sea-surface temperatures and of surface wind speeds. The atmospheric parameters are essentially eliminated by the use of multi-channel/multi-frequency approaches. Quoted accuracies for the geophysical parameters extracted from SMMR data are (Thomas, 1981):

- Sea-surface temperature ± 1.5 K;

- Surface windspeed ± 2.0 ms^{-1}.

The spatial resolution for this kind of data is very low and small-scale variations in atmospheric conditions will tend to be smoothed out over the instantaneous field of view. In principle one might attempt to account for large-scale variations in atmospheric conditions by using simultaneous meteorological data. However, because of the poor spatial resolution and corresponding frequent large-area coverage, SMMR data are principally used for large scale meteorological, climatological and oceanographic studies in remote regions of the Earth, such as the Pacific Ocean and Antarctica, where weather stations are very few and far between. Thus the method adopted for atmospheric corrections to SMMR data is, essentially, a version of option 5 outlined in Section 8.3, using a multi-channel or, more specifically, a multi-frequency approach. The instrument has five frequency channels and for each frequency there are two possible planes of polarization of the microwave radiation, so that there are ten radiance values for each pixel. Atmospheric corrections are made using a model atmosphere; the required atmospheric and geophysical parameters being determined by fitting the ten radiance values for each pixel to the model.

Thus, once one has set up the atmospheric model, the values of the geophysical parameters can be extracted in a routine manner from the satellite data. Since the ten radiance values are obtained for each pixel, one is able effectively to recalculate the parameters in the atmospheric model for each pixel. Thus, although one is working within the constraints of a given atmospheric model one is able to make allowance for variations in atmospheric conditions on a pixel-by-pixel basis, which of course is exactly the right scale on which to handle the variations in atmospheric conditions when processing scanner data. There is no need to have any *in situ* data for calibration purposes, though at the initial stages of handling the data from any given system and sensor one will need *in situ* data for validation purposes and, possibly, for refining the atmospheric model used. In other words, with SMMR data it is possible to use remote sensing as it really should be used, that is to extract environmental data completely from the remote sensing data without having to appeal to any simultaneous *in situ* calibration data from another source.

The fact that ten radiance values are obtained per pixel with the SMMR means that the processing of SMMR data is a much more self-contained problem than the handling of data from optical and infrared scanners. In general, optical and infrared scanners flown on satellites provide fewer radiance values per pixel than the SMMR, see Table 8-2, and the opportunity for making corrections on a pixel-by-pixel basis to data from optical and infrared scanners is consequently more restricted.

8.4.3 Thermal-infrared scanner data

Data from a thermal-infrared scanner can be processed to yield the value of the radiance leaving the surface of the Earth. The intensity of the radiation leaving the Earth, at a given wavelength, will depend on both the temperature of the surface and the emissivity of the surface. The emissivity of the sea is known to be very close to unity, about 0.98 in fact. Over the land the value of the emissivity varies widely from one surface to another. Consequently one has to regard both the emissivity and the temperature of the land surface or land cover as unknowns that have to be determined

Table 8.2   Radiance values per pixel

Scanner	No. of channels
LANDSAT MSS	4 (occasionally 5)
LANDSAT TM	7
AVHRR	5 (sometimes 4)
METEOSAT	3
SMMR	10

either exclusively from the scanner data or from the scanner data plus supplementary data.

An intermediate step, before one considers the effect of atmospheric conditions, is to calculate the brightness temperature; this is the equivalent black-body temperature of the radiation that is incident on the satellite. Radiation which arrives at a satellite is incident on detectors which produce voltages in response to this radiation incident upon them. The voltages produced are then digitized to create the digital numbers that are transmitted back to Earth. The data generated by a satellite-borne thermal-infrared scanner are received at a ground receiving station as a stream of digital numbers — often as 8-bit numbers, but 10-bit numbers in the case of the AVHRR. The first task to be undertaken in the quantitative use of thermal-infrared data from a scanner on a satellite is to calibrate the data by working backwards through these steps, using both pre-flight and in-flight calibration data to calculate the intensity of the radiation incident on the instrument. Assuming that the energy distribution of the incident radiation is that of black-body radiation one can calculate the temperature corresponding to that radiation by inverting the Planck radiation formula. This temperature is known as the brightness temperature. The accuracy that can be attained in determining the brightness temperature depends on the internal consistency and stability of the scanner and on the accuracy with which it is able to be calibrated. Brightness temperatures can be determined to an accuracy in the region of 0.1 K from the AVHRR on the TIROS-N series of satellites.

The basic procedures involved in converting the raw thermal-infrared scanner data into brightness temperatures are well-established. The two important steps are

1. The digital data have to be converted into radiances in absolute units; and

2. The Planck radiation formula has to be inverted to obtain the brightness temperature.

To these we should add a third step which is important in practice if one tries to process all reasonable scenes and not just the very small proportion of completely cloud-free scenes; this step is a fairly easy and accurate method and must be used for

3. The identification and removal of cloud-covered areas in a scene.

Let us first consider step 3, since there is no point in processing data corresponding to areas of cloud that are obscuring the surface of the sea. Various methods are available for identifying cloudy areas. Many scenes are just solid cloud and can be rejected immediately by visual inspection. There would seem to be no need to improve on this simple technique in these cases. Scenes which are partially cloudy are much more difficult to handle. One method would be to use an interactive system and to outline the areas of cloud manually, using a light pen to draw round the cloudy areas, and to have appropriate software organized to reject those areas. Alternatively, one can try to establish automatic methods for the identification of clouds. Several such methods are described and illustrated for AVHRR data by Saunders (1982). These include the use of the visible channel with a visible threshold, a local uniformity method, and a histogram method; the use of the 3.7 μm channel data for cloud identification is also considered.

With regard to step 1, the calibration of the thermal-infrared channels of the AVHRR is achieved using two calibration sources which are viewed by the scanner between scanning the surface of the Earth; these two sources comprise a black body target of measured temperature on board the spacecraft and a view of deep space. Taking the scanner data, together with pre-flight calibration data supplied by NOAA, the conversion of the digital data into radiances can be achieved. The inversion of the Planck distribution function to obtain the brightness temperature is then a standard mathematical operation. If T denotes the mean target temperature then the radiance $L(\kappa, T)$ in a spectral channel detected by a detector is computed from

$$L(\kappa, \bar{T}) = \sum_{i=1}^{n} B(\kappa_i, \bar{T}) \hat{\phi}(\kappa_i) \, \Delta\kappa_i \qquad (8\text{-}32)$$

where $B(\kappa_i, \bar{T})$ is the Planck distribution function, $\hat{\phi}(\kappa_i)$ is the normalized spectral response function of the detector, n denotes the number of discrete wave numbers within the spectral window at which the response of the detector was measured during the preflight investigation, $\Delta\kappa_i$ is the width of the ith interval within which the response function was measured. NOAA Technical Memorandum NESS 107 (Lauritson *et al.*, 1979) supplies 60 values of the normalised response function with specified values of $\Delta\kappa_i$. Using these data, the standard radiance $L(\kappa, \bar{T})$ can be evaluated from equation 8-32. Strictly speaking the response of a detector may not be a linear function of the radiance; it is necessary to make a suitable correction for this (Singh and Warren, 1983). The radiance $L^*(\kappa)$ in a spectral channel detected by a detector for an instantaneous field of view (IFOV) in a scan line is related to the brightness temperature T_b by

$$L^*(\kappa) = \sum_{i=1}^{n} B(\kappa_i, T_b) \hat{\phi}(\kappa_i) \, \Delta\kappa_i \qquad (8\text{-}33)$$

where κ refers to the spectral channel centred around the wavenumber κ. In principle, this relation can be solved for the temperature T_b, but notice that T_b occurs in the integrand on the right hand side. Thus one does not try to solve this equation explicitly for T_b. One selects a range of T_b which is appropriate to the sea-surface temperatures that are likely to be encountered; since the response function of the detector is known, the value of $L^*(\kappa)$ can be computed for various trial values of T_b until the observed value of $L^*(\kappa)$ is reproduced. However, this is time-consuming and once this has been done for a number of pixels one can use the calculated brightness temperatures to determine the coefficients in an empirical relationship of the form

$$\ln(L^*(\kappa)) = \alpha + \frac{\beta}{T_b} \qquad (8\text{-}34)$$

where α and β are parameters which depend on the selected range of T_b and the absolute value of T_b. This formula can then be used to calculate the brightness temperatures very quickly for the bulk of the scene (see Singh and Warren, 1983).

Having calculated the brightness temperature T_b for the whole of the area of sea surface in the scene, the problem of calculating the atmospheric correction then has to be tackled since the objective of using thermal-infrared scanner data is to obtain information about the temperature or the emissivity of the surface of the land or sea. It is probably fair to say that the problem of atmospheric corrections has been studied fairly extensively in relation to the sea but has received little attention for data obtained from land surface areas. As seen in Section 8.3 there is upwelling atmospheric radiance, downwelling atmospheric radiance and radiation from space in addition to the surface radiance that is required to be determined. Moreover, as indicated in Section 8.2, radiation propagating through the atmosphere will be attenuated. These effects are considered separately here.

Figure 8-7 shows the contributions from sea-surface emission, reflected solar radiation and upwelling and downwelling emission for the 3.7 μm channel of the AVHRR; the units of radiance are T.R.U. (where 1 T.R.U. = 1 mW/m^2 sr cm^{-1}). It can be seen that at this wavelength the intensity of the reflected solar radiation is in the region of $10 - 15\%$ of that of the radiation emitted from the surface, while the atmospheric emission is very small. Figure 8-8 shows data for the 11 μm channel of the AVHRR. It can be seen that the reflected radiation is of little importance but that the atmospheric emission, though small, is not entirely negligible.

The data in Figures 8-7 and 8-8 are given for varying values of the atmospheric transmittance. The actual value of the atmospheric transmittance or of the atmospheric attenuation at the time that the scanner data were collected is needed to make quantitative corrections to a given set of thermal-infrared scanner data. Of the three attenuation mechanisms mentioned in Section 8.3.3, namely Rayleigh (molecular) scattering, aerosol scattering and aerosol absorption by gases, it is the absorption by gases that is the important mechanism in the thermal infrared, where water vapour,

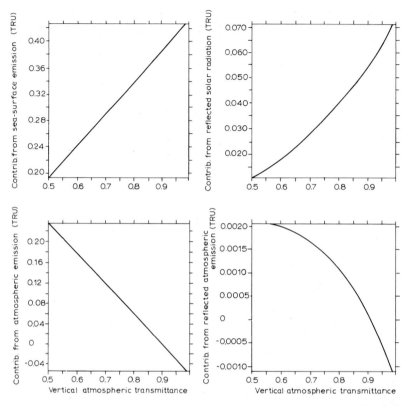

Figure 8-7 Various components of the satellite-recorded radiance in the 3·7 μm channel of the AVHRR (Singh and Warren, 1983)

carbon dioxide and ozone are the principal atmospheric absorbers and emitters (see Figure 8-6). To calculate the correction that must be applied to the brightness temperature to give the temperature of the surface of the sea it is necessary to know the concentrations of these substances in the column of atmosphere between the satellite and the target area on the surface of the sea. However, these concentrations, especially that of water vapour, are very variable and the pressure and humidity profiles in this atmospheric column would need to be known in addition for these corrections to be calculated. Only about 13% of the energy emitted at the surface of the Earth at infra-red wavelengths escapes from the top of the atmosphere to space. The remainder is absorbed by the atmosphere. However, although the average transmittance is so low, the variation with wavelength is relatively large. There are three spectral intervals, or "windows", in which the vertical transmittance may be as high as 90%; these windows are from 3 – 5 μm, 7 – 8 μm and 9.5 – 14 μm, see Figure 8-6. Of these three windows the 3 – 5 μm window is the most nearly transparent; unfortunately, this window has proved to be of less practical

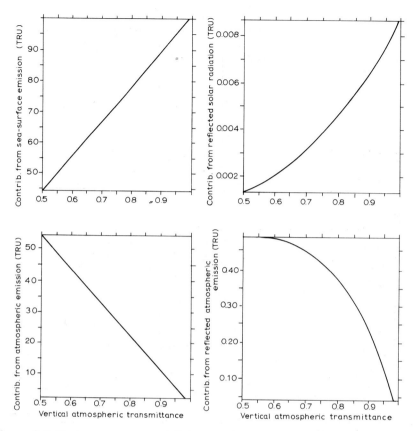

Figure 8-8 Various components of the satellite-recorded radiance in the 11 μm channel of the AVHRR (Singh and Warren, 1983)

value than the others because of the large amount of reflected solar radiation at this wavelength. Values of the atmospheric correction, expressed in Kelvin and calculated for a number of standard atmospheres, are given in Table 8-3 (Phulpin and Deschamps, 1980). Calibration of AVHRR data using results from buoys, ships and low-flying aircraft carrying scanners has been undertaken by various workers, (Bernstein, 1982; Bowers *et al.*, 1982; Singh and Warren, 1983; Singh *et al.*, 1985), and other workers have used data from geosynchronous satellites (Maul, 1981).

Assuming that the atmospheric conditions are the same throughout a given scene, i.e. that the atmosphere is horizontally stratified, the magnitude of the atmospheric correction will be the same, though difficult to calculate precisely, throughout the scene. Thus relative differences between the sea surface temperature at two points within the scene can be determined quite accurately, often to better than 0.1 K, using the corresponding difference between the brightness temperatures at those two points. But the

Table 8.3 Atmospheric attentuation, T_a, for various standard atmospheres

Atmosphere	channel	H$_2$O lines	CO$_2$	O$_3$	N$_2$ cont.	H$_2$O cont.
Tropical	1[b]	1·31	0·46	0	0·22	0·41
	2[c]	0·86	0·30	0	0	3·44
	3[d]	2·80	0·55	0	0	4·89
Midlatitude summer	1	0·95	0·42	0	0·20	0·26
	2	0·59	0·27	0	0	1·61
	3	2·05	0·49	0	0	2·39
Midlatitude winter	1	0·38	0·36	0	0·17	0·09
	2	0·21	0·21	0	0	0·23
	3	0·83	0·39	0	0	0·35
Subarctic summer	1	0·85	0·42	0	0·20	0·24
	2	0·53	0·26	0	0	1·23
	3	1·87	0·47	0	0	1·86
US Standard	1	0·76	0·45	0	0·22	0·19
	2	0·48	0·29	0	0	0·78
	3	1·76	0·53	0	0	1·20

(b) 1 refers to the 3·7μm channel
(c) 2 refers to the 11 μm channel
(d) 3 refers to the 12 μm channel

(after Phulphin and Deschamps, 1980)

differences between the actual sea surface temperatures and the brightness temperatures are substantially greater than this; they may be as large as 1 K and are often considerably greater than this. For the data in Figure 8-9(a) the correction that was needed to the brightness temperature was only of the order of 0.1 K, but for Figures 8-9(b) and 8-9(c) it was about 3.0 K and 3.3 K, respectively. A mean value of 1.8 K for the difference over a twelve month period is quoted by Tournier (1978), see Figure 8-10, while a value varying from 2.6 to 0.1 K for data in the vicinity of Fiji is quoted by McKenzie and Nisbet (1982) and values varying from 0.74 to 3.58 K for the Irish Sea are quoted by Bowers *et al.*, (1982).

It is possible to show that the magnitude of the atmospheric correction varies very considerably from one point to another within a given AVHRR scene and this causes further complication. These results are shown in Figure 8-11 in which the difference between the brightness temperature and the sea-surface temperature has been calculated using radiosonde data to give the atmospheric parameters at five stations around the coastline of the U.K. The calculations were performed using the method of Weinreb and Hill (1980) which incorporates a version of LOWTRAN. For a sea-surface

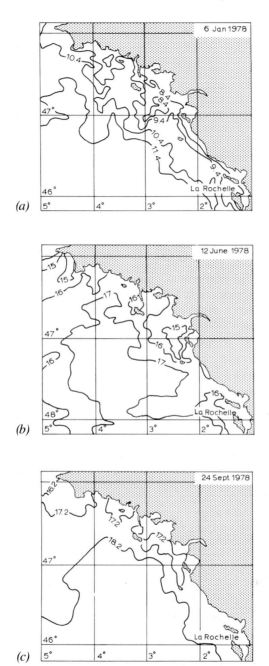

Figure 8-9 Sea surface temperatures off the French coast determined from NOAA-5 VHRR data with subsequent correction with in situ data (a) 6 January 1978, (b) 12 June 1978 and (c) 24 September 1978 (Cassanet, 1981)

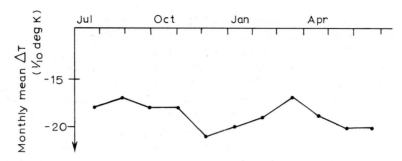

Figure 8-10 Differences between satellite-derived brightness temperatures and measurements from ships for period from summer 1975 to summer 1976 (Tournier, 1978)

temperature of 15°C (288 K) the correction varies from about 0.5 K at some stations to about 1.5 K at other stations. Thus we conclude that method 3 outlined in Section 8.3 is not appropriate to thermal-infrared data and that if method 4 outlined in Section 8.3 is to be used it is necessary to obtain sufficient simultaneous meteorological data to allow local variations in atmospheric conditions to be taken properly into account.

For some years NOAA have been using AVHRR data to produce sea-surface temperature maps on a global scale at very low spatial resolution with waters near to the U.S.A. being covered at higher spatial resolution. With the older version of the AVHRR, in which there were only two infrared channels, the option that has been described as method 4 in Section

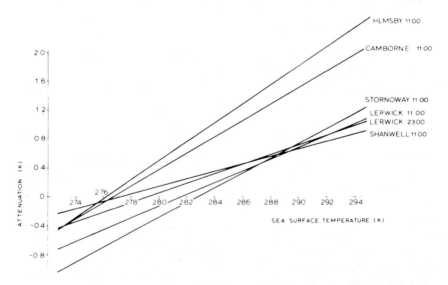

Figure 8-11 Atmospheric attenuation versus sea surface temperature calculated using radiosonde data for five stations around the UK (from Callison and Cracknell, 1984)

8.3 was used. That is, an atmospheric model was assumed and the parameters within the model were determined using data from the soundings obtained from other instruments flown on the same satellite. However, the soundings give data obtained from a coarser grid and therefore not exactly co-registered with all the scanner data. With the introduction of the split window AVHRR/2 with two bands at 10.3 – 11.3 μm and 11.5 – 12.5 μm in place of a single band at 10.5 – 11.5 μm NOAA have changed to one of the techniques outlined as method 5 in Section 8.3, namely the use of a multi-channel approach to the elimination of atmospheric effects. This approach seems to have been suggested first by Anding and Kauth (1970) and since then it has been applied by a number of workers (Rao, *et al.*, 1972; Prabhakara *et al.*, 1974; Sidran, 1980); this has the advantage that all local variations in atmospheric conditions are eliminated on a pixel-by-pixel basis. The original suggestion of Anding and Kauth was to use two bands, one between 7 μm and 9 μm and the other, on the other side of the ozone band, between 10 μm and 12 μm. They argued that since the same physical processes are responsible for absorption in both of these wave bands the effect in one ought to be proportional to the effect in the other. Therefore one measurement in each wavelength interval could be used to eliminate the effect of the atmosphere. A more formal justification of the technique can be obtained by manipulation of the radiative transfer equation. This involves making several assumptions, namely

- That the magnitude of the atmospheric correction to the brightness temperature is fairly small;

- That one only includes the effect of water vapour; and

- That the transmittance of the atmosphere is a linear function of the water vapour content;

for details see, for example, Singh and Warren (1983). The sea-surface temperature T_s is then written in the form

$$T_s = e_0 + e_1 T_B(\kappa_1) + e_2 T_B(\kappa_2) \tag{8-35}$$

where κ_1 and κ_2 refer to the two spectral channels and the parameters e_0, e_1 and e_2 are determined by a multi-variate regression analysis. For the AVHRR this method was first used for the two infrared channels on the older versions of the instrument, i.e. with wavelength ranges 3.55 – 3.93 μm and 10.5 – 11.5 μm. This has subsequently been extended to a three-band approach using the three bands having ranges 3.55 – 3.93 μm, 10.3 – 11.3 μm and 11.5 – 12.5 μm (for example Phulpin and Deschamps, 1980; Bowers *et al.*, 1982). In tests of the multi-channel approach carried out by Bowers *et al.* with a data buoy recording sea-surface temperatures in the Irish Sea it was concluded that using the two-channel technique, with the split channel at 10–12 μm, a rms error of rather less than 1 K was found.

With the three-channel algorithm which can only be used with night-time data because of the significant quantity of reflected solar radiation in the 3.55 – 3.93 μm channel during the day, an rms error of about 0.3 K was found. The data from the instruments on the buoy used by Bowers *et al.* were recorded on the buoy and recovered by ship afterwards. An alternative method of recovery of sea-surface temperature data from buoys using the ARGOS data collection system has been developed (Cracknell and Singh, 1981); an important advantage of the use of the ARGOS system is that the sea-surface temperature measurement data from the buoy are downlinked from the satellite in the same data stream as the AVHRR data and so are available in real time for calibration and checking of the calculated sea-surface temperatures.

Approximate values of the (rms) error in the sea-surface temperature calculated from satellite data using methods 1 – 5 are given in Table 8-4; the errors have not, so far, been reduced to the level of the instrumental noise of the AVHRR thermal-infrared channel(s).

8.5 *Visible wavelength scanners*

8.5.1 Principles of atmospheric corrections

The various atmospheric corrections to data from optical scanners flown on satellites have been discussed in general terms in Section 8.3. As mentioned, for visible wavelengths the absorption by molecules involves ozone only. The ozone contribution is relatively small and not too difficult to calculate to the accuracy required. The most important contributions are the Rayleigh and aerosol scattering, both of which are large, particularly towards the blue end of the optical spectrum. Moreover, since it is reflected radiation that is of concern, light that reaches the satellite by a variety of other paths has to be considered in addition to the sunlight reflected from the target area, see Section 8.3.2 and Figure 8-3. It is, therefore, not surprising that the formulation of the radiative transfer equation for radiation at visible wavelengths is rather different from the approach used in Section 8.4 for microwave and thermal-infrared wavelengths. In this

Table 8.4   *RMS errors in sea surface temperatures derived from AVHRR data*

Method	1	2	3	4	5
rms error (K)	2	0·1 – 0·2	1·2	≤ 1	2/30·9[a] 2/30·3[b]

Note: methods 1–5 are described at the beginning of Section 8·3
[a] *two-channel data*
[b] *three-channel data*

section the term "visible" is taken to include, by implication, the near-infrared wavelengths as well, up to about 1.1 μm or so in wavelength.

There are various situations to be considered regarding atmospheric corrections to data in the visible spectrum. Land-based applications are distinguished from aquatic applications. In the past there has been a considerable amount of work done on atmospheric corrections to satellite data with reference to applications to water bodies such as inland lakes/lochs, rivers, estuaries and coastal waters. This has been based on the use of LANDSAT data (Alfoldi and Munday, 1978; Ahern and Murphy, 1978; Aranuvachapun and LeBlond, 1981; Klemas *et al.*, 1974; Maracci, 1978*a*, 1978*b*; Munday and Alfoldi, 1979; MacFarlane and Robinson, 1984) and CZCS data (Singh and Cracknell, 1979; Cracknell and Singh, 1980; Sturm, 1981, 1983; Singh, *et al.*, 1983, 1985).

In aquatic applications it is the water-leaving radiance that is principally extracted from the satellite data. From the water-leaving radiance one attempts to determine the distribution and concentrations of chlorophyll and suspended sediment. Ideally the objective is to be able to do this without any need for simultaneous *in situ* data for calibration of the remotely sensed data. The atmospheric effects are by no means small at visible wavelengths, particularly towards the blue end of the spectrum and, as a consequence, the accuracy that can be obtained in the extraction of geophysical parameters from visible-channel satellite data without simultaneous *in situ* calibration data is limited.

For land-based applications, rather than aquatic applications, there has been virtually no work done on considering the quantitative importance of atmospheric effects on classification work using satellite data over land. Some estimates are given based principally on CZCS data, though making use of LANDSAT data. It will be recalled, (see Table 3-1), that the NIMBUS-7 CZCS includes bands of shorter wavelength than those of the LANDSAT MSS. The shorter the wavelength then the greater is the importance of the atmospheric effects; for example for band 1 of the CZCS $L_W(\lambda)$ the water-leaving radiance constitutes only about 20-30% of $L(\lambda)$, the satellite-received radiance, see Table 8-5. For bands 2 and 3 $L_W(\lambda)$, considered as a fraction of $L(\lambda)$, does increase, see Table 8-5, but it is then much smaller in band 4. The sharp drop in the fractional contribution of $L_W(\lambda)$ in band 4 is associated with the fact that the reflectivity of the water at the red and near-infrared wavelengths is very small. $L_W(\lambda)$ values in band 4 are only about one tenth of the $L_W(\lambda)$ values in band 3. A similar situation can be expected to apply to LANDSAT MSS data since bands 4 and 5 of the LANDSAT MSS (at $0.5 - 0.6$ μm and $0.6 - 0.7$ μm respectively) overlap with bands 2, 3 and 4 of the CZCS. For band 6 and band 7 of LANDSAT MSS data the water-leaving radiance is negligibly small. From the values of $L_W(\lambda)/L(\lambda)$ it can be seen that in considering aquatic data and the extraction of values of physical or biological parameters of water masses

Table 8.5   Water-leaving radiance as a proportion of satellite-received radiance

CZCS — over water			Equivalent LANDSAT bands	
Band	$\lambda(\mu m)$	$L_W(\lambda)/L(\lambda)$	$\lambda(\mu m)$	Band
1	$0 \cdot 433 - 0 \cdot 453$	$\sim 20 - 30\%$		
2	$0 \cdot 510 - 0 \cdot 530$	$\sim 40\%$	$0 \cdot 5 - 0 \cdot 6$	4
3	$0 \cdot 540 - 0 \cdot 560$	$\sim 50\%$		
4	$0 \cdot 660 - 0 \cdot 680$	$\sim 10 - 20\%$	$0 \cdot 6 - 0 \cdot 7$	5
5	$0 \cdot 700 - 0 \cdot 800$	0	$0 \cdot 7 - 0 \cdot 8$	6

Note: $L_W(\lambda)=0$ is assumed for band 5 of CZCS over clear water in the calculation of the atmospheric corrections

- It is absolutely essential to take into account the atmospheric effects; and

- Only modest errors in calculating the atmospheric corrections will lead to very great errors in derived $L_W(\lambda)$ values.

Another way of looking at the quoted results is to say that the total atmospheric contribution to the satellite data at optical wavelengths is well over 50% and may approach 80% or 90% of the radiance received at the satellite. This is very much in contrast with the case of thermal-infrared scanners for which, as was indicated in the previous section, the corrections that have to be made to the satellite-received radiance to produce the Earth-leaving radiance are of the order of 1 or 2%. It is accordingly clear that the application of atmospheric corrections to optical scanner data to recover quantitative values of the Earth-leaving radiance is very much more difficult and therefore needs to be performed much more carefully than in the thermal-infrared case.

It is not possible to perform a calculation directly for the land to obtain results like those given for the sea in Table 8-5. There are two reasons for this. The first is that the method used for calculating the aerosol contribution to the atmospheric effects cannot be applied over the land. The second is that in bands 1 – 4 of the CZCS the signal saturates the detectors over land. However, attempts have been made to estimate $L_L(\lambda)/L(\lambda)$, that is the land-leaving radiance as a fraction of the satellite received radiance. These estimates were made from results calculated for typical CZCS data over areas of water, the assumption being that the same atmospheric contribution applies over adjacent land as well, which is not at all likely. The satellite-received radiance over land was then estimated by using ratios of satellite-received radiance over land to satellite-received radiance over water obtained from typical LANDSAT MSS data in bands 4 and 5; the value of this ratio was found to be about 2:1. Values of $L_L(\lambda)/L(\lambda)$ were obtained from these estimates. The results which apply to the CZCS bands are presented in Table 8-6. As in Table 8-5 estimates from bands 2 and 3 of the CZCS can be used to apply to band 4 of the LANDSAT MSS and the estimate from band 4 of the CZCS can be used to apply to band 5 of the LANDSAT MSS, see Table 8-6.

In land-based studies the atmospheric corrections have in the past been regarded as less important than in aquatic applications for two main reasons:

1. The intensity of the radiation $L_L(\lambda)$ leaving the surface of the land is larger than that leaving the water so that, proportionately, the atmospheric effects are less significant in land-based studies than in aquatic studies which utilize optical scanner data.

2. The data used in land-based applications have been derived mostly from the LANDSAT MSS and accordingly involve the longer wavelength part of the visible spectrum where the atmospheric effects are less important than for the blue end of the spectrum.

As far as 1 is concerned, the values of $L_L(\lambda)/L(\lambda)$ in Table 8-6 compared with the values of $L_W(\lambda)/L(\lambda)$ in Table 8-5 support this. As far as 2 is concerned, this is also supported by the estimates in Table 8-6; it can be seen that for the LANDSAT MSS the atmospheric correction for bands 4, 5 and 6 varies between about 10% and about 40%. Although no calculations have been done for wavelengths corresponding to band 7 of the LANDSAT MSS it would be expected that for many land-based applications, especially those involving healthy vegetation, the atmospheric correction would be similar to that in band 6, i.e. of the order of 10% or less. However, with regard to 2, once extensive use begins to be made of the blue band of the Thematic Mapper data from LANDSAT-4 and -5, the atmospheric corrections are going to become more important, see Table 3-1.

Table 8.6 Estimation of land-leaving radiance

CZCS Band	1	2	3	4	5
WATER					
$L(\lambda)$ (digital nos)	123	142	164	134	10
$L_W(\lambda)/L(\lambda)$	0·25	0·4	0·5	0·15	—
$L_{atm}(\lambda)$ (digital nos)	92	85	82	113	10
LAND					
$L(\lambda)$ (digital nos)*	248	284	328	268	100
$L_L(\lambda)$ (digital nos)	156	199	246	155	90
$L_L(\lambda)/L_L(\lambda)$	0·63	0·70	0·75	0·58	0·90

$L(\lambda)$ is estimated for bands 1 to 4 (see text)

There seems to be virtually no work described in the literature on the question of determining the consequences of the atmospheric effects for the classification of multi-spectral data for land-cover applications. It would seem that there is a rather urgent need for some work to be carried out on this subject.

In Section 8.5.2 the various options that have been outlined will be considered from the point of view both of the data and of the nature of the application in which it is proposed to use the data.

8.5.2 Corrections to visible-band data

It has already been noted that for the visible channels the atmospheric contribution to the radiance received at a satellite forms a very much greater percentage of the radiance leaving the target area than is the case for the thermal infrared. Thus any attempt to make atmospheric corrections to visible-channel data using method 3 outlined in Section 8.3 with a model atmosphere with values of the parameters determined only by the geographical location and by the time of the year is likely to be very inaccurate. To obtain good results with a model atmosphere with simultaneous meteorological data, as in method 4 outlined in Section 8.3, it would be likely that the meteorological data would be required over a much finer network of points than is usually available. It therefore seems that, although some of the less important contributions to the atmospheric correction for the visible channels may be estimated reasonably well using model atmospheres or sparse meteorological data, in order to achieve the best values for the atmospheric corrections in the visible channels one must expect to have to use a multi-channel approach for some of the contributions to the atmospheric correction. Of the various mechanisms mentioned in Section 8.3.3 it is the aerosol scattering which is the most difficult contribution to determine. We shall summarize, briefly, the approach that has been used for making atmospheric corrections to CZCS data (Cracknell and Singh, 1981; Singh *et al*, 1983) and discuss briefly the problems associated with using atmospherically-corrected radiances in algorithms for the calculation of water quality parameters such as chlorophyll and suspended sediment concentrations.

The discussion in this section will be given, by implication at least, in terms of large open areas of water and will therefore be relevant to marine and coastal applications. The question of relating discussions of atmospheric corrections of data from water surfaces to land-use applications was considered in some detail in Section 8.5.1. This was based on the extrapolation to the land area of an atmospheric correction calculated over water; such an approach can be regarded as reasonably valid provided the atmospheric conditions do not vary drastically over the distance involved in the extrapolation. For such land areas the alternative to such an extrapolation is to revert to using a model atmosphere with a very sparse network of meteorological sounding data.

The radiance $L(\lambda)$ received by a sensor in a spectral channel with wavelength λ can be expressed as

$$L(\lambda) = \{L_W(\lambda) + L_g(\lambda)\}T(\lambda) + L_p^A(\lambda) + L_p^R(\lambda) \qquad (8\text{-}36)$$

where

$L_W(\lambda)$ = water-leaving radiance

$L_g(\lambda) = L_s(\lambda) + L_d(\lambda)$ with

$L_s(\lambda)$ = Sun glitter radiance

$L_d(\lambda)$ = diffused sky glitter radiance

$T(\lambda)$ = proper transmittance, i.e. the transmittance from the target area to the sensor

$L_p^A(\lambda)$ = aerosol path radiance and

$L_p^R(\lambda)$ = Rayleigh (molecular) scattering path radiance.

In order to extract the water-leaving radiance, which is the quantity that contains the useful information about the target area on the surface of the sea, all other quantities appearing in equation 8-36 must be evaluated. Methods exist for calculating $L_s(\lambda)$, $L_p^R(\lambda)$ and $L_p^A(\lambda)$ directly, see for example Sturm (1981), but the calculation of $L_p^A(\lambda)$ is more difficult. The calculation of $L_p^A(\lambda)$, the aerosol path radiance, has been considered by Gordon (1978). In his approach it was argued that if the water-leaving radiance in the 670 nm wavelength band from the target area corresponding to the darkest pixel in the scene is assumed to be zero, then the radiance detected by the remote sensor in that wavelength channel is due to atmospheric scattering and the glitter only. Then the aerosol path radiance for this particular wavelength can be evaluated. Moreover, the aerosol path radiance for any other wavelength can be expressed in terms of the now known aerosol path radiance in the 670 nm channel. This method is known as the "darkest pixel" method but it does have some problems. Thus, the darkest pixel in the extract chosen from the scene may not be the darkest pixel in the whole scene; unless the choice of the darkest pixel can be verified each and every time, one cannot be sure of having found the correct pixel. It is evident that there is some degree of arbitrariness in defining the darkest pixel in a scene. In spite of this, the darkest pixel-approach of Gordon, which has subsequently been used by many other workers, gave real hope for the quantitative applicability of CZCS data for marine and coastal waters. A way around this arbitrariness was proposed by Smith and Wilson (1980); they introduced an iteration technique as follows. Assuming that the subsurface upwelling radiance $L_{ss}(670)$ for the 670 nm band is zero then the subsurface upwelling radiance in all the other visible channels can be calculated. Now suppose that $L_{ss}(670)$ is non-zero. Then $L_{ss}(670)$ is computed using the relation

$$\frac{L_{ss}(443)}{L_{ss}(670)} = A \left(\frac{L_{ss}(443)}{L_{ss}(670)} \right)^B \tag{8-37}$$

where the values of $A = 12.06$ and $B = 1.661$ are given by Smith and Wilson (1980). This new value of $L_{ss}(670)$ is then used to calculate the new values of the subsurface upwelling radiance in the other channels. These new values can then be inserted into the recurrence relation 8-37 to give a new value of $L_{ss}(670)$. This whole iterative process is continued until a pre-determined accuracy is attained. It has to be admitted, however, that there is no guarantee that the values of A and B quoted above have any universal validity. The final quantity that has to be evaluated before one can use equation 8-36 to determine the water-leaving radiance is $L(\lambda)$, the radiance detected at the scanner on the satellite. Before we go any further with the discussion of the extraction of this radiance from the digital numbers in the raw data stream from the CZCS, we would like to pursue the question of the calculation of the aerosol scattering a little further.

The recurrence relation due to Smith and Wilson which we have quoted is not entirely satisfactory. In particular, there is no reason to assume that the values actually chosen for the constants A and B in equation 8-37 have any universal validity. Consequently we have also recently been studying other ways to try to determine the aerosol path length, still purely from the satellite data itself. If we suppose that the glitter can be assimilated into the Rayleigh and aerosol path radiances then equation 8-36 reduces to

$$L(\lambda) = L_p^R(\lambda) + L_p^A(\lambda) + L_W(\lambda)T(\lambda) \tag{8-38}$$

The assumption of $L_W(750) = 0$ was suggested by Gordon (1978) before the launch of NIMBUS-7, but when CZCS data for clear waters actually became available $L_W(670) = 0$ was assumed instead. It can then be argued that one can consider the possibility of treating the water-leaving radiance in both these channels as being zero. Under this assumption equation 8-38 then reduces to

$$L(\lambda) = L_p^R(\lambda, \tau_A(\lambda)) + L_p^A(\lambda, \tau_A(\lambda)) \tag{8-39}$$

for both of the wavelengths 670 nm and 750 nm. It is $\tau_A(\lambda)$ that is unknown and it is assumed to be a function of wavelength. One can make the conventional assumption that $\tau_A(\lambda)$ can be written in the form already given

$$\tau_A(\lambda) = A\lambda^{-B} \tag{8-40}$$

where A and B are considered to be constants independent of wavelength for a given type of aerosol but will be different for different types of aerosol. It is then possible to regard A and B as the two unknowns in the two equations of the form 8-39 and thence to determine the values of A and B. This approach has been tested for two CZCS scenes, one of the Bristol Channel of 14 March 1980 and the other of the Irish Sea of 15 March 1981.

The criterion which was used for identifying the clear water (or darkest pixel) was that the digital data in the 670 nm and 750 nm were the smallest of all the digital data in these channels. The smallest values of the digital data found in these channels for the first of the scenes studied were 85 and 6 in the 670 nm and 750 nm channels, respectively. It should be understood that the inherent striping present in the CZCS data has been removed beforehand. Also, if there is any bright object then there is an undershooting present in the CZCS electronics when it scans immediately beyond a bright object, thereby giving a small value of the digital output. Such pixels should be avoided while finding the darkest pixels.

Expressions for the aerosol path length obtained by Singh from the CZCS data for these two scenes were

$$\tau_A(\lambda) = (0.158 \pm 0.005)\lambda^{-(0.6 \pm 0.1)} \tag{8-41}$$

for the first scene (14 March 1980) and

$$\tau_A(\lambda) = (0.325 \pm 0.009)\lambda^{-(0.5 \pm 0.1)} \tag{8-42}$$

for the second scene (15 March 1980). Notice that the values of A and B do vary from one scene to another. We are not postulating a universal pair of values of A and B that one can apply to all scenes; the indications seem to be that one needs to calculate values of A and B for each scene from the data in that scene itself. It is important, however, to stress that since we are only using two channels and determining A and B from the data for these two channels there are no spare degrees of freedom that would enable one to test the basic assumption, namely that $\tau_A(\lambda)$ can be written in the form of equation 8-40 where the Ångström exponent B is taken to be a constant. An aircraft-flown scanner that uses more than two channels in the near-infrared would provide evidence to establish whether there may be some departures from the Ångström relation, even though the relation is very widely used in practice. Further refinements which ought also to be taken into consideration include

- Making allowance for the variation of the refractive index of sea water as a function of wavelength of the radiation;

- The variation in the temperature and salinity of the sea water; and

- The fact that different research workers have chosen different values for the refractive index of sea water.

The effect of taking these factors into consideration is discussed at some length by Singh *et al.* (1983); in that paper estimates are made of the magnitudes of the errors associated with these factors.

Finally, mention must be made of the problems associated with determining the absolute values of the radiance incident at a satellite from the digital numbers in the output data stream. The radiation falls on the scanner, is filtered, falls on to the detectors and the voltage generated by each detector is digitized to produce the output digital data. To convert these digital numbers back into voltages one makes use of the fact that every so often during flight a voltage staircase is applied to the digitizing electronics and the output from this is also transmitted in the data stream. Having converted the digital numbers into voltages, it is then necessary to convert these voltages back to incident radiance values. For some scanners flown on satellites this has to be done using pre-flight instrument calibration data, but this is clearly not satisfactory as the performance of the instruments is likely to change from its pre-flight performance. For the visible channels of the CZCS inflight calibration is done by making use of the standard lamps on board the spacecraft. Every so often the scanner views these standard lamps and the digital data output from this are also included in the data stream. For the thermal-infrared channel the calibration is done with a black body on board the spacecraft and the measured temperature of the black body is also included in the data stream from the instrument. Other problems that have to be taken into account include the gain setting that can be applied to channels 1 to 4 and the tilting of the scanner's mirror to avoid sunglint.

In the case of the thermal-infrared wavelengths the rule for extracting the temperature, albeit only a brightness temperature, was based on a fundamental physical formula, namely the Planck radiation formula (see Section 8.4.3). In the case of the extraction of marine physical or biological parameters from CZCS data the situation is much less straightforward; it would be very difficult to obtain from first principles a relationship between the radiance and the concentration of suspended sediment or of chlorophyll in the water. Consequently empirical relationships are used and these most commonly take the form

$$M = A(r_{ij})^B \tag{8-43}$$

where M is the value of the marine parameter and r_{ij} is the ratio of the reflectances, or radiances, at the two wavelengths λ_i and λ_j. The marine parameters commonly studied in this manner are the pigment concentration C (chlorophyll-a and phaeophytin-a, mg m^{-3}), the total suspended load S (dry mass, g m^{-3}) and the diffuse attenuation coefficient K (m^{-1}) for a given λ. Because of the wide range of variations in the marine parameters, the natural trend was to perform regression analyses with log-transformed data. If a significant linear relationship is found, an algorithm of the above-mentioned type is obtained. Table 8-7 contains a list of such algorithms proposed by diverse authors (Sathyendranath and Morel, 1983). The overall impression that one obtains from this table is that various ratios can be used and that the values of the coefficients A and B vary widely from one set of data to another. It seems clear that more work is needed both

- To check that the results of atmospheric corrections are in accord with experimental data on water-leaving radiances;

Table 8.7 Some values of parameters given by different workers for algorithms for chlorophyll and suspended sediment concentrations from CZCS data

r_{ij}	A	B	N	r^2
$M = Chl\ a + Pheo\ a\ (mg\ m^{-3})$				
L_{443}/L_{550}	0·776	−1·329	55	0·91
L_{443}/L_{520}	0·551	−1·806	55	0·87
L_{520}/L_{550}	1·694	−4·449	55	0·91
L_{520}/L_{670}	43·85	−1·372	55	0·88
L_{440}/L_{550}	0·54	−1·13	7	0·96
L_{440}/L_{550}	0·505	−1·269	21	0·978
L_{440}/L_{520}	0·415	−1·795	21	0·941
L_{520}/L_{550}	0·843	−3·975	21	0·941
R_{440}/R_{560}	1·92	−1·80	67	0·97
L_{443}/L_{550}	0·783	−2·12		0·94
L_{443}/L_{520}	0·483	−3·08		0·88
L_{520}/L_{550}	2·009	−5·93		0·95
L_{443}/L_{550}	2·45	−3·89	6	0·61
L_{443}/L_{550}	1·13	−1·71	454	
L_{443}/L_{550}	1·216	−2·589		
$M = Total\ suspended\ particles\ (g\ m^{-3})$				
L_{440}/L_{550}	0·4	−0·88	9	0·92
L_{440}/L_{520}	0·33	−1·09	9	0·94
L_{520}/L_{550}	0·76	−4·38	9	0·77
L_{443}/L_{550}	0·24	−0·98		0·86
L_{520}/L_{550}	0·45	−3·30		0·86
L_{520}/L_{670}	5·30	−1·04		0·85

(data gathered by Sathyendranath and Morel, 1983)

- To seek to understand better the applicability of the algorithms and, hopefully, determine how to establish the values of the coefficients in the algorithm for scenes for which simultaneous *in situ* data are not available.

The above discussion has been expressed in terms of the CZCS. Similar treatments can be developed for other scanners that have optical channels, such as the MSS on LANDSAT-1, -2, -3, the Thematic Mapper on LANDSAT-4, the scanner on SPOT, etc., but the details will be different. In some instances at least the opportunity for in-flight calibration may be more restricted than it is with the CZCS. Some research has been done on the calibration of MSS data from LANDSAT-1, -2, -3 (Ahern and Murphy, 1978; Alfoldi and Munday, 1978; MacFarlane and Robinson, 1984; Munday and Alfoldi, 1979); there is much to be done with the Thematic Mapper and SPOT.

9 Image processing

9.1 Introduction

A very high proportion of the data used in remote sensing exists and is used in the form of images, each image containing a very great deal of information. Image processing involves the manipulation of images

- To extract information;

- To emphasize or de-emphasize certain aspects of the information contained in the image; or

- To perform statistical or other analyses to extract non-image information.

Image processing may therefore be regarded as a branch of Information Technology. Some of the simpler operations of image processing will be familiar from everyday life; for example, the idea of contrast enhancement will be familiar from one's experience of photography or of television viewing.

A distinction needs to be made between analogue and digital methods in image processing. The distinction is not absolute, however, and many sets of image data are transformed from digital to analogue form or vice versa at some stage in their history. The term analogue image is applied to an image which exists on a photographic film, or as a video signal from a television broadcast transmission or video magnetic tape. The processing of an optical image is carried out using lenses, mirrors, stops or other optical components and the processing of a video signal is carried out using the appropriate electronic circuitry. The term digital image is applied to an array or matrix of numbers where each element of the array corresponds to an element of the image (a picture element or pixel). Digital image data may therefore be processed conveniently with a computing system; this may be a general-purpose mainframe computer, minicomputer or microcomputer or it may be a purpose-built image processing system with its own dedicated computing power. For a human observer, however, the digital image is nearly always converted to analogue form using a chosen "grey scale" relating the numerical value of the element in the array to the density on a photographic material or to the brightness of a spot on a screen. It is interesting to make comparisons between optical and digital image processing techniques. For example optical systems are often simple in concept, relatively cheap and easy to set up and they provide very rapid processing of images. Also, from the point of view of introducing image processing, optical methods often demonstrate some of the basic principles involved rather more clearly than could be done digitally. There are many formal texts on image processing and some of these are given in Appendix I.

This chapter is concerned only with image processing relating to remote sensing problems. In the vast majority of cases remotely sensed image data are processed digitally and not optically, although in most cases the final output products for the user are presented in analogue form. One notable recent exception was the case of synthetic aperture radar, SAR, data from SEASAT which was first optically processed or "survey processed" to give "quick look" images; digital processing of the SAR data, which is especially time-consuming, was then carried out later only on useful scenes selected by inspection of the survey-processed quick-look images. Traditionally a great deal of use has been made of aerial photography in cartographic work. In the past the techniques involved in the incorporation of photogrammetric results into cartographic databases have been manual or photographic, i.e. analogue, techniques. However, with the steady expansion of the use of digital databases in cartography there is now also a great increase in the use of digital photogrammetric techniques.

Table 9-1 summarizes the common digital image processing operations that are likely to be performed on data input as an array of numbers from some computer-compatible medium. There are corresponding operations to most of these processes for the case of processing analogue images.

9.2 The format of a digital image

A digital image consists of an array of numbers which may be square but is quite commonly rectangular. Digital images are likely to have been

Table 9.1 Some common digital image processing operations

Histogram generation	Multispectral classification
Contrast enhancement	Neighbourhood averaging and filtering
Histogram equalization	De-striping
Histogram specification	Edge enhancement
Density slicing	Principal components
Classification	Fourier transforms
Band ratios	High-pass and low-pass filtering

generated directly by a scanner or created from an analogue image by a densitometer. Each row of the array or matrix will normally correspond to one scan line. The numbers are almost always integers and it is common to work with one byte, i.e. one 8-bit number, for each element of the array. This 8-bit number, which must therefore lie within the range $0 - 255$, denotes the intensity, or grey scale value, associated with one element (a picture element or pixel) of the image. To produce a display of the image on a screen or to produce a hard copy on a photographic medium, the intensity corresponding to a given element of the array is mapped to the brightness assigned to the spot on the screen or the shade of grey assigned to the photographic medium for that picture element or pixel. There is, of course, the possibility of producing a "positive" or "negative" image from a given array of digital data, though the question of which is positive and which is negative is largely a matter of definition. A picture with good contrast will be obtained if the intensities associated with the pixels in the image are well distributed over the range from 0 to 255, that is for a histogram such as that shown in Figure 9-1.

9.3 Image generation

At the outset a hard-copy image corresponding to the raw data is likely to be required. Hard-copy images may also be required after some processing operations have been performed. A photofacsimile machine, filmwriter, laser beam or fibre-optic recorder, etc. is used to generate a high-quality image. As an alternative to purpose-built devices and where lower quality is acceptable, or where rapid examination is required and probably financial constraints apply, conventional computer printers may be

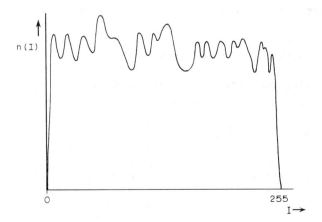

Figure 9-1 Sketch of histogram of n(I) number of pixels with intensity I, against I for an image with good contrast

used. Images generated in this way have been available for some time and are often produced for demonstration or for social purposes. It is possible to use a lineprinter to produce a simple image which will give a quick look at the data and enable the user to find out what range of values are present. The ability to see both the basic shapes in the scene represented by the digital data and the range of these numbers is a prerequisite to the processing of an image. A high quality grey level picture of the image is desirable for visual representation but may be time consuming and costly to produce; it also gives little information on the digital values themselves although providing information on their relative variation. On the other hand, the printing out of the actual digital data as the numbers themselves may be quickly done, but the resulting large quantity of computer printout cannot readily give an overall awareness of an image both in terms of physical size and ability to distinguish levels because one can only look at a small part at one time. For rapid comprehension one needs an output form that is physically manageable, reproduces the digital values and is visually meaningful. One convenient approach is to assign each of the 256 grey levels to one of the possible 64 lineprinter characters. The benefit of printing out the values as a single character is that one can see three times the number of columns on the same piece of computer printout and there is relatively little loss of grey scale resolution.

Broadly speaking, there are two types of computer output printers, namely those using formed characters (as in a lineprinter) and those using a dot matrix. Each requires different treatment to generate a grey scale. An example of a grey scale with 7 levels is illustrated in Figure 9-2.

Formed character printers can be used with single characters to represent the numerical values of the elements of the array of digital data but such printers are best used by overprinting with a faint ribbon to generate a grey scale.

The overprinting technique is a time consuming operation, but since lineprinters are fast devices this generally does not matter. Dot matrix printers are usually much slower than lineprinters and do not always use ribbons; some use heat sensitive paper or ink-jets. Multi-access terminals are often of matrix type and often use 132 column paper.

While a dot matrix printer is usually set up to print ASCII characters there is no need for it to operate in this mode. The print head of a typical dot matrix printer (an EPSON is used as an example) has nine pins or wires

Figure 9-2 7-level greyscale formed from lineprinter characters (the first character is blank)

arranged vertically above one another. There are 480 printing positions across the line. When printing characters each character is normally formed by a 9 by 5 array generated by five successive positions of the print head along the line. Adjacent characters are separated by a blank position and so there are 480/6 = 80 characters per line. Upper-case letters are generated within the upper 7 by 5 part of the array, the lowest two rows being reserved for the descenders of commas, semi-colons and lower-case letters. In the normal character mode there will be spaces between adjacent characters and between adjacent lines. In the bit image printing mode the spaces between lines are eliminated and there are no characters, just a succession of adjacent positions of the print head all the way along each line. The printing of dots for each position of the print head is controlled by the number (from 0 to 255) that is used as the input data for printing in that position. Only 8 of the 9 pins are used and they are controlled by the "1"s and "0"s in the data. If a bit is set the print head fires and if a bit is not set the print head does not fire for the wire in question. For example if the input data are $34(22_H)$[1]or $80(50_H)$ the patterns of dots printed in Figure 9-3 will be produced.

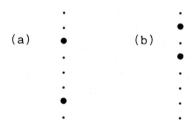

Figure 9-3 Dot patterns printed by (a) 22_H and (b) 50_H where ● represents a printed dot and . represents a blank; the most significant bit is at the top (Cracknell et al., 1982b)

By the separate specification of the input data to the print head for each position of the print head on each line in the image one can produce any required "half-tone" picture. The problem is to assemble the image data in the form required by such a printer.

As an example of the construction of a grey scale one might choose to work with a basic 4 by 4 array of dots to represent each pixel in the image data. This gives us 480/4 = 120 pixels in each scan line of the image. With an array of 4 by 4 dots one can construct a 16-level grey scale and one example is shown in Figure 9-4.

One can then print an image with 120 pixels on each line and each printed line has a height of 8 dots and corresponds to two scan lines of the image data.

---

[1]Subscript H denotes hexadecimal notation.

Image processing

Pixel intensity	0–15	16–31	32–47	48–63	64–79	80–95	96–111
Printed pattern	●●●● ●●●● ●●●● ●●●●	●●●● ●.●● ●●●● ●●●●	●●●● ●.●● ●●.● ●●●●	●●●. ●.●● ●●.● ●●●●	●●●. ●.●● ●●.● .●●●	●.●● ●●.● ●.●. .●●●	.●●. ●.●● ●●.● .●●.

112–127	128–143	144–159	160–175	176–191	192–207	208–223	224–239	240–255
.●●. ●.●● ●..● .●●.	●.●. .●.. ●.●. .●.●	..●. .●.. ●.●● .●.●	..●. .●.● ●.●● .●..	.... .●.● ●.●. .●..	.... .●.● ●.●.	.... .●.● ..●.	.... .●.. ..●.	....●. ....

Figure 9-4 One example of a dot pattern for a 16-level grey scale for use with a dot matrix printer (Cracknell et al., 1982b)

9.4 Density slicing

Density slicing is considered at this stage because it is very closely related to the discussion given to image generation. The use of a 7-level grey scale or of a 16-level grey scale, as considered in Section 9.3, were really special cases of density slicing. The objective of density slicing is to reduce the number of grey levels in an image by redistributing the levels into a given number of specified slices to facilitate visualisation of features in the image. With the 16-level grey scale illustrated in Figure 9-5 the range of intensities from 0 – 255, corresponding to 256 grey levels, has been divided into 16 ranges with each range having 16 grey levels assigned to it.

This illustrates a mathematical approach aimed at producing an image that is capable of being printed on a dot-matrix printer but associated also with the fact that the human eye is incapable of discriminating more than about 16 different shades of grey in a black and white image. However, there may be some good reasons, associated with the physical origins of the digital data, for using a different number of ranges (or "bins") and for using unequal ranges for the different bins. For example, in a LANDSAT MSS image the intensity received at the scanner in band 7 from an area of the surface of the Earth that is covered by water will be extremely low. Supposing, for the sake of argument, that all these intensities were lower than 10. Then a convenient and very simple density slice would be to assign one grey value, say white, to all pixels with intensities in the range 0 – 9 and a second value, say black, to all pixels with intensities in the range 10 – 255. In this way a simple map that distinguished land areas (black) from water areas (white) would be obtained. The scene represented by this image could then be thought of as having been classified into areas of land and areas of water. This is a very simple two-level slice and classification scheme and more complicated classification schemes can be envisaged. For example, water could perhaps be classified into sea water and fresh water

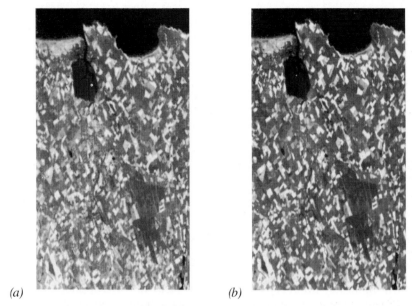

(a)   *(b)*

Figure 9-5 (a) 16-level grey scale, (b) 256-level grey scale

and the land could be classified into agricultural land, urban land areas and forestry. Further subdivisions, both of the water and of the land area, can be envisaged; however, the chance of achieving a very detailed classification on the basis of a single LANDSAT band is not very good.

9.5 Image processing programs

In designing an image processing system, a device for displaying an image on a screen or writing it on a photographic medium would usually be set up so as to produce a picture with good contrast when there is a good distribution, or full utilization, of the whole range of intensities from 0 to 255. If the intensities are all clustered together in one part of the range, as for example in Figure 9-6, the image will have very low contrast; the contrast could be restored by a hardware control, like the contrast control on a television set. However, it is often more convenient to keep the hardware setting fixed and to perform operations such as contrast enhancement digitally before the final display or hard copy is produced. For images with histograms such as that shown in Figure 9-6 the intensities can all be scaled by software to produce a histogram like that in Figure 9-1 before producing a display or hard copy. An important component of any image processing software package is therefore bound to be a program for generating histograms from the digital data. Although histograms are important in

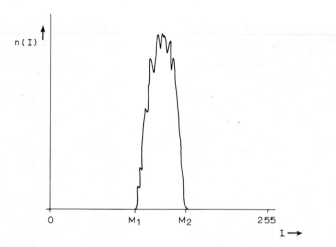

Figure 9-6 Sketch of histogram for a mid-grey image with very poor contrast

their own right, they probably have their greatest use when applied to image enhancement techniques such as contrast stretching, density slicing and histogram specification. In addition to constructing a histogram for a complete image, there may be some reason for constructing a histogram either for a small area extracted from the image or for a single scan line.

9.6 Image enhancement

It is convenient to think in terms of a transfer function which maps the intensities in the original image into intensities in a transformed image with an improved contrast. Suppose that $I(x,y)$ denotes the intensity associated with a pixel labelled by x (the column number or pixel number) and y (the row number or scan-line number) in the original image and that $I'(x,y)$ denotes the intensity associated with the same pixel in the transformed image. Then

$$I'(x,y) = T(I)I(x,y) \qquad (9\text{-}1)$$

where $T(I)$ is the transfer function. A transfer function might have the appearance shown in Figure 9-7; it is a function of the intensity in the original image, but not of the pixel coordinates x and y. This particular function would be suitable for stretching the contrast of an image for which the histogram was "bunched" between the values M_1 and M_2, see Figure 9-6. It has the effect of stretching the histogram out much more evenly over the whole range from 0 to 255 to give a histogram with an appearance such as that of Figure 9-1. The main problem is to decide on the form to be used for the transfer function $T(I)$. It is very common to use a simple linear

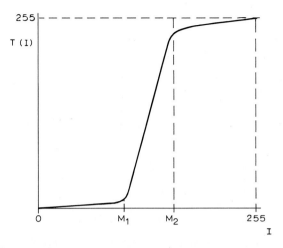

Figure 9-7 Transfer function for contrast enhancement of the image with the histogram shown in Figure 9-6

stretch as shown in Figure 9-8. A function that resembles more closely the transformation function in Figure 9-7 is shown in Figure 9-9 where $T(I)$ is made up of a set of straight lines joining the points that correspond to the points of inflection in Figure 9-7. The points of inflection of the curve can be regarded as parameters that must be specified for any given digital image; suitable values of these parameters can be determined by inspection of the histogram for that image.

The desirability of producing good contrast by having a histogram of the

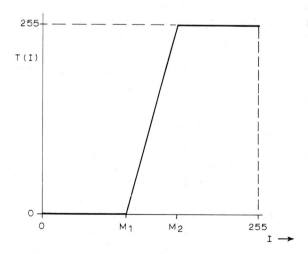

Figure 9-8 Transfer function for a linear contrast stretch for an image with the histogram shown in Figure 9-6

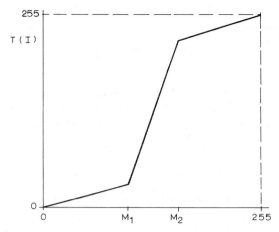

Figure 9-9 A transfer function consisting of three straight-line segments

form of Figure 9-1 has already been mentioned. Contrast stretching can be regarded as an attempt to produce an enhanced image with a histogram of the form of Figure 9-1. A completely flat histogram may, however, be produced in a rather more systematic way known as "histogram equalization". If I ($=0,255$) represents the grey levels in the image to be enhanced then the transformation $J = T(I)$ will produce a new level J for every level I in the original image. Continuous variables i and j replace the discrete variables I and J for the grey levels. The probability density functions $p(i)$ and $p(j)$ are considered. A continuous transfer function $T(i)$, can then be thought of in place of the transfer function $T(I)$, where

$$j = T(i) \tag{9-2}$$

The graphs of $p_i(i)$ against i and of $p_j(j)$ against j are simply the histograms of the original image and of the transformed image, respectively. $p_j(j)$ and $p_i(i)$ are related to each other by

$$p_j(j) \; dj = p_i(i) \; di \tag{9-3}$$

so that

$$p_j(j) = p_i(i) \frac{di}{dj} \tag{9-4}$$

To achieve histogram equalization a special transfer function has to be chosen so that $p_j(j) = 1$ for all j. Therefore $p_i(i)di/dj=1$ or $dj/di=p_i(i)$. Integrating this

$$j = \int_0^i p_i(w) \; dw \tag{9-5}$$

and comparing this with the definition

$$j = T(i) \tag{9-6}$$

of the transfer function, $T(i)$, it can be seen that

$$T(i) = \int_0^i p_i(w) \, dw \qquad (9\text{-}7)$$

That is, equation 9-7 defines the particular transfer function that will achieve histogram equalization. This can be illustrated analytically for a simple example. Consider $p_i(i)$ shown in Figure 9-10 where

$$\left.\begin{array}{ll} p_i(i) = 4i & 0 \le i \le \tfrac{1}{2} \\ = 4(1-i) & \tfrac{1}{2} \le i \le 1 \end{array}\right\} \qquad (9\text{-}8)$$

The transfer function to achieve histogram equalization will then be given by

$$\left.\begin{array}{ll} T(i) = 2i^2 & 0 \le i \le \tfrac{1}{2} \\ = -1 + 4i - 2i^2 & \tfrac{1}{2} \le i \le 1 \end{array}\right\} \qquad (9\text{-}9)$$

This function is plotted in Figure 9-11 and the transformed histogram is shown in Figure 9-12.

With a digital image, a discrete distribution and not a continuous distribution is being considered with i and j replaced by the discrete variables I and J. An approximation can be made to equation 9-7

$$J = T(I) = \sum_{J=0}^{J=I} PI(J) \qquad (9\text{-}10)$$

where the original histogram is simply a plot of $PI(I)$ versus I in this notation. The histogram for the transformed grey levels J generated in this way will be the analogue, for the discrete case, of the uniform histogram produced by equation 9-7; it will not be quite uniform, however, because discrete rather than continuous variables are being dealt with. The degree

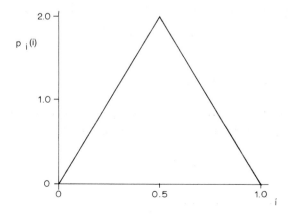

Figure 9-10 Schematic histogram defined by equation 9-8

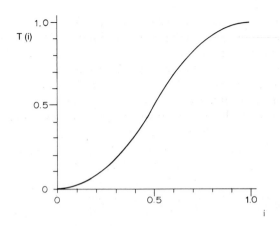

Figure 9-11 Transfer function defined in equation 9-9

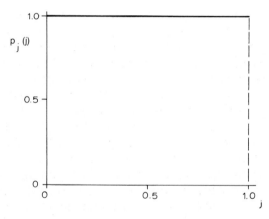

Figure 9-12 Transformed histogram obtained from $p_i(i)$ in Figure 9-10 using the transfer function shown in Figure 9-11

of uniformity does, however, increase as the number of grey levels is increased.

The last transformation has been concerned with transforming the histogram of the image under consideration to produce a histogram for which $p_j(j)$ was simply a constant. It is, of course, also possible to define a transfer function to produce a histogram corresponding to some other given function, such as a Gaussian function, a Lorentzian function etc., rather than just a constant.

The enhancement considered so far has been contrast enhancement. That is, pixel intensities have been manipulated simply on the basis of the intensities themselves and irrespective of their position in the image. Another type of enhancement is edge enhancement. This enhancement is used to "sharpen" an image by making clearer the positions of boundaries between (moderately) large features in the image. There are a variety of reasons why edges of a feature in a digital image may not be particularly sharp. To carry out edge enhancement a boundary has first to be identified and then appropriate action taken to enhance the boundary. In order to identify the boundary the differences between the intensity of a pixel and those of adjacent pixels are examined by studying the gradient or rate of change of intensity with change in position. Near a boundary the value of the gradient will be large while some distance from a boundary the gradient will be small. The enhancement can then be achieved, for example, by increasing the brightness (pixel intensity) by adding to the original (unenhanced) pixel intensity an amount proportional to the gradient of the intensity. This is done not just for those pixels at the boundary but for all the pixels in the entire image. For pixels far away from the boundary the effect of adding this in will be very small but for pixels near the boundary it will increase the intensity very significantly. The effect is illustrated in Figure 9-13.

Before concluding this section we ought briefly to mention the concept of image smoothing. We have considered image enhancement so far in this section in terms of contrast enhancement or edge enhancement, that is trying to improve the quality of an image, in terms of accentuating items of detail, or to improve the aesthetic acceptability of an image. In other words, we have been thinking in terms of trying to find small details in an image and to accentuate or emphasize them in some way. The concept of smoothing an image may then, at first sight, seem an odd thing to want to do and to be counterproductive in terms of enhancing the usefulness or acceptability of an image. However, smoothing becomes important when small details in an image are noise rather than useful information. Indeed, many images do contain noise and it is important, desirable and, sometimes, necessary to remove the noise. There are a number of different ways of doing this. These include

- Neighbourhood averaging;

- Low-pass filtering; and

- Averaging of multiple images.

Neighbourhood averaging is a real-space technique; each pixel intensity is replaced by some chosen form of weighted average of the intensity of that pixel and the intensities of a set of neighbouring pixels. Low-pass filtering is, essentially, a reciprocal space, or inverse space or Fourier space technique. One takes a Fourier transform of the image and then applies some chosen form of filter to the Fourier transform where the filter allows through the low-(spatial)-frequency components and blocks the high-(spatial)-frequency components. It has been assumed, in what has been said so far, that one is considering random noise in the image. If the

(a)

(b)

Figure 9-13 Illustration of edge enhancement (a) original image and (b) enhanced image

noise is of a periodic type, such as for instance the six-line striping found in much of the old LANDSAT MSS data, then the idea of neighbourhood averaging can be applied on a line-by-line, rather than on a pixel-by-pixel, basis. However, alternatively, using a Fourier-space filtering technique is particularly appropriate; in this case the six-line striping will give rise, in the Fourier transform or spectrum of the image, to a very strong component at the particular spatial frequency corresponding to the six lines in the direction of the path of travel of the scanner. By identifying this component in the Fourier transform spectrum of the image, removing just this component and then taking the inverse Fourier transform, one has a very good method of removing the striping from the image. The third method, namely the averaging of multiple images, is only applicable in certain rather special circumstances. First of all one must have a number of images which are assumed to be identical except for the presence of the random noise. This means that the images must be co-registered very precisely to one another. Then what is done for any given pixel, specified in position by coordinates x and y, is to take an average of the intensities of the pixels in the position x and y in each of the co-registered images. Since the noise is randomly distributed in the various images this averaging will tend to reduce the effect of the noise and enhance the useful information. In practice the situation in which this kind of multiple-image averaging is important is in synthetic aperture radar images. A synthetic aperture radar image will contain speckle and this is frequently overcome by using sub-sets of the raw data to generate a small number of independent images which will all contain speckle. These images will automatically be co-registered and they will all contain speckle, but the speckle in each image will be independent of that in each of the other images; consequently the averaging of these multiple images will reduce the speckle and enhance the useful information very considerably.

9.7 Multi-spectral images

So far it has been supposed that a digital image consists of one array of numbers and that each element of the array of numbers corresponds to a single picture element (pixel). However, while a single array of numbers can quite adequately represent a monochrome image, if a coloured picture is to be represented by a digital set of data it is necessary to specify one number for each of the three primary colours for each pixel. The elements of the array may be thought of as consisting not just of scalar numbers but of three-dimensional vectors or of ordered sets of three numbers. An alternative representation would be to consider three arrays of numbers corresponding to three co-registered images associated with the three primary colours. The number of intensities associated with a given pixel need not, however, be restricted to three. Digital data from multi-spectral scanners with N channels can be regarded either as an array where the elements of the array are N-dimensional vectors or as a set of N co-registered arrays. If N is equal to 3 then each of the three components of the image can be associated with one of the primary colours and a coloured image may accordingly be generated. There is, of course, no reason why the coloured image generated in this manner should faithfully reproduce the true-colour balance of the original object or scene. Data from three channels of a multi-spectral scanner are assigned to the three colour layers of a colour film or to the three colour guns of a colour television monitor. While the image will be produced in colour, it will almost certainly be a false-colour (composite) image as it does not include all the wavebands naturally assigned to each colour.

In the case of data from the multi-spectral scanner on the LANDSAT series of satellites, for example, the conventional approach is to assign bands 1, 2 and 4 to the colours blue, green and red, respectively. In this way terrain with healthy vegetation, which has a high intensity of reflected radiation in the near-infrared (band 4) and very little reflection in the yellow-orange (band 1) will appear red, while areas of water, with very little reflected radiation in the near-infrared (band 4), will appear blue. Although this particular assignment of LANDSAT bands to colours does not produce an image with the original colour balance of the scene it is, nevertheless, widely used.

Contrast enhancement may be applied separately to each band. By allowing for separate enhancement of each band, i.e. making separate choices of M_1 and M_2 for each band in Figure 9-6 it is clear that an enormous number of different shades of colour may be produced in the image. Operations applied to the individual bands separately, such as varying the contrast stretch, may be very valuable in extracting information from the digital data. However, to carry out such manipulations easily and quickly it is more or less essential to have access to an interactive digital image processing system.

About ten years ago image processing systems for handling remotely sensed data were very rare. An image processing system consists basically of a set of computer hardware with some specialized peripheral devices, together with a suite of software to perform the necessary image processing and display operations. In some systems some of the operations that are performed most frequently may be carried out by hard-wired logic rather than by software. Ten years ago it was quite likely that one might try to build an image processing system for serious research work by acquiring the necessary hardware, most of it as standard computer hardware items, together with a frame store and display and write the necessary software for oneself. Over the last ten years the situation has changed in several ways. At the top end of the market there are now a large number of proprietary turnkey image processing systems that are readily available, are competitively priced and well supported in terms of maintenance of the hardware and software on a worldwide basis. At the lower end of the market the recent developments in terms of microcomputers means that one can now put together a reasonable image processing facility with a microcomputer. Indeed microcomputer-based systems are approaching the situation in which they can be used as serious research machines in lieu of some of the very expensive proprietary systems.

The idea of classifying a monochrome image has already been mentioned in Section 9.4. A multi-spectral image provides the possibility of obtaining a more refined classification than is possible with a single spectral band. Different surfaces have different spectral reflecting properties. If a variety of surfaces are considered, the reflected intensities in the different spectral bands will generally be different for a given illumination. This is illustrated in the sketches shown in Figure 9-14.

Consider the case of three spectral bands for example. If the data from three bands are used to produce a false-colour composite image then a surface with a given spectral signature will be associated with a particular colour in the image. As previously mentioned, in a conventional LANDSAT MSS false-colour composite healthy vegetation will appear red, water will appear blue and urban areas will appear grey. A classification of a LANDSAT scene could be carried out on the basis of a visual interpretation of the shade of colour and, indeed, a great deal of environmental work has been carried out on this basis.

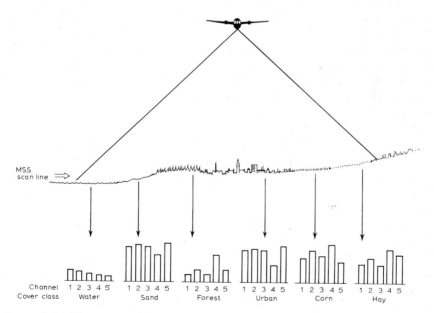

Figure 9-14 Illustration of the variation of the spectra of different land cover classes in multi-spectral scanner data: band 1, blue; band 2, green; band 3, red; band 4, near-infrared; band 5, thermal-infrared (adapted from Lillesand and Kiefer, 1987)

As an alternative to a visual classification of a false-colour image, a classification could be carried out digitally within a computer. Apart from any advantages associated with the use of a computer rather than a human interpreting colours, there is the added advantage of the digital processing approach that it can handle more than three bands simultaneously and therefore, hopefully, obtain a more sensitive and accurate classification. Figure 9-15 represents a three-dimensional space in which the coordinates along the three axes are the intensities in the three spectral bands under consideration, e.g. bands 1, 2 and 3 of the LANDSAT multi-spectral scanner.

For any pixel in the scene a point defined by the values of the intensities in the three spectral bands for that pixel can be located on this diagram. Ideally, all the pixels corresponding to a given type of land cover would then be expected to be represented by a single point in this diagram, but in practice they will be clustered close together. However, these points may form a cluster that is distinct from, and quite widely separated from, the clusters corresponding to pixels associated with other types of land cover. Therefore, provided the pixels in the scene do group themselves into well-defined clusters that are quite clearly separated from one another, Figure 9-15 can be used as the basis of a classification scheme for the scene. Each cluster will be identified with a certain land cover; this may be done either from a study of a portion of the scene selected as a training area or from experience. By specifying the coordinates, i.e. the intensities in the three bands, for each cluster, and by specifying the size and land cover of each cluster, it should then be possible to assign any given pixel to the appropriate class. If "training data" are used from a portion of the scene to be classified quite good accuracy of classification is obtainable. But any attempt to classify either a sequence of scenes obtained from a given area on a variety of different dates, or a set of scenes from different areas, with

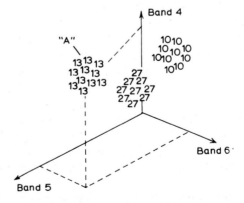

Figure 9-15 Sketch to illustrate the use of cluster diagrams in a three-dimensional feature space for three-band image data

the same training data, should only be made with extreme caution. This is because, as mentioned, the intensity of the radiation received, in a given spectral band, at a remote sensing platform, will depend not only on the reflecting properties of the surface of the land or sea but also on the illumination of the surface and on the atmospheric conditions. If satellite-received radiances, or aircraft-received radiances, are used without conversion to surface reflectances or normalized surface-leaving radiances (where the normalization takes account of the variation in solar illumination of the surface) the classification using a diagram of the form of Figure 9-15 will not be immediately transferable from one scene to another.

If more than three spectral bands are available, the computer programs used to implement the classification that is illustrated in Figure 9-15 can readily be generalized to work in an N-dimensional space where N is the number of bands to be used.

9.8 *Principal components*

The idea of the principal components transformation follows on from the discussion of multi-spectral images in the previous section. It has been pointed out that in general a multi-spectral image contains more information than an image in a single band. It has also been shown that it is possible to extract information from several bands by carrying out some kind of multi-spectral classification procedure. When information from three bands is combined and represented in a single false-colour image there is the possibility of a much greater number of distinguishable shades of the whole spectrum of colours instead of only having different shades of grey. The information in a multi-spectral image may be distributed fairly uniformly among the various bands. The principal components transformation can be regarded as a transformation of the axes in a diagram such as Figure 9-15. This principal components transformation may be carried out with the intention of creating a new set of bands in which the information content is not distributed fairly uniformly among the bands but rather distributed so that the information content is concentrated as much as possible into a small number of transformed bands. After carrying out the principal components transformation, the maximum information content of the image will be found in the first principal component or transformed band, and decreasing amounts in subsequent transformed bands. If each transformed band is viewed as a monochrome image the first principal component will contain very high contrast, while the last principal component will show virtually no contrast and be an almost uniform shade of grey.

The principal components transformation was originally proposed by Hotelling (1933), but has been subsequently developed by a number of authors. The origins of the transformation were in the statistical treatment of data in psychological problems, long before the possibility of its application to the treatment of image data in general or of remote sensing image data in particular was appreciated. Since many people find the idea of principal components difficult to understand in the multi-spectral image case, a brief summary is given of what is involved in the rather simpler psychological case that was originally considered by Hotelling.

Consider a set of n variables, $x_1, x_2,...,x_n$, attached to each individual of a population. In Hotelling's original discussion he considered the scores obtained by school children in reading and arithmetic tests. It is usual to expect that the variables x_i will be correlated. Now consider the possibility of the existence of a more fundamental set of independent variables, perhaps fewer in number than the x_i, which determine the values the x_i will take. These variables are denoted by $y_1, y_2,...$ to establish a set of relations of the form

$$x_i = f_i(y_1, y_2...) \tag{9-11}$$

where $i = 1,2,...n$. The quantities y_i are then called *components* of the complex depicted by the tests. Now consider only normally distributed systems of components which have zero correlations and unit variances; this may be summarized conveniently by writing

$$E(y_i y_j) = \delta_{ij} \tag{9-12}$$

where δ_{ij} is the Kronecker delta. The argument is simplified by supposing that the functions f_i are linear functions of the components so that

$$x_i = \sum_{j=1}^{n} a_{ij} y_j \tag{9-13}$$

Assuming that the matrix **A**, with elements a_{ij}, is non-singular this relationship can be inverted and the components y_k written in terms of the variables x_i

$$y_k = \sum_{i=1}^{n} b_{ki} x_i \tag{9-14}$$

If r_{ik} is the correlation between x_i and x_k, i.e.

$$r_{ik} = E(x_i x_k) \tag{9-15}$$

r_{ik} has the property that $r_{ik} = 1$ if $i = k$ and for the remaining values $r_{ik} = r_{ki}$. (Hotelling worked in terms of standard measures z_i obtained by taking the deviation of each x_i from its mean, $\bar{x}_i$, and dividing by its standard deviation, σ_i, to simplify the formulation.) These conditions on the r_{ik} are insufficient to enable the coefficients a_{ij} in the transformation 9-13 to be determined completely. In other words the choice of components is not completely determined and one has in fact an infinite degree of freedom in choosing the components y_i (or the coefficients a_{ij} or the coefficients b_{ij}). There are various methods which enable the coefficents a_{ij} to be determined completely. For example, just sufficient of the coefficients might be set equal to zero to ensure that the remainder are determined, but not overdetermined, by the conditions imposed by the properties of r_{ik}. The method adopted by Hotelling was to say that from the infinite number of possible modes of resolving the variables into components, the following method is chosen: begin with a component y_1 whose contribution to the variances of the variables x_i is the greatest possible, then take a second component y_2 that is independent of y_1 and whose contribution to the variances is also as great as possible, subject to its own independence of y_1; similarly y_3 is chosen to maximize the variance, subject to y_3 being independent of y_1 and y_2. The remaining components are determined in a similar manner, with the total not exceeding n in number, although some of the components may be neglected because their contribution to the total variance is small. This is described as the method of *principal components*. The detailed derivation of the formulae that enable one to determine the principal components, which involves the use of Lagrange's undetermined multipliers, is relatively straightforward though slightly tedious and is given by Hotelling (1933).

The example given by Hotelling is worth mentioning. He took some data from the results of tests of 140 school children and considered correlations for reading speed ($i=1$), reading power ($i=2$), arithmetic speed ($i=3$) and arithmetic power ($i=4$). The values of the correlations are shown in Table 9-2. The result of transforming into principal components is given in Table 9-3. The first principal component seems to measure general ability, while the second principal component seems to measure a difference between arithmetical ability on the one hand and reading ability on the other. Together, these account for 83% of the variance. An additional 13% of the variance, corresponding to the third principal component, seems to be a matter of speed versus deliberation. The remaining contribution to the variance, associated with the fourth principal component, is negligible. One would gain a very good idea of the information content of the results of the tests on the school children from

Table 9.2 Correlations for Hotelling's original example

i \ j	1	2	3	4
1	1·000	0·698	0·264	0·081
2	0·698	1·000	−0·061	0·092
3	0·264	−0·061	1·000	0·594
4	0·081	0·092	0·594	1·000

Table 9.3 Principal components for Hotelling's original example

	Y_1	Y_2	Y_3	Y_4	Totals
Root	1·846	1·465	0·521	0·167	3·999
% of total variance	46·5	36·5	13	4	100
Reading speed	0·818	−0·438	−0·292	0·240	
Reading power	0·695	−0·620	0·288	−0·229	
Arithmetic speed	0·608	−0·674	−0·376	−0·193	
Arithmetic power	0·578	0·660	0·459	0·143	

only the first two principal components; the third component contains relatively little information and the fourth component almost nothing at all.

The approach can now be re-formulated in terms of multi-spectral images. Suppose that $I_i(p,q)$ denotes the intensity, in the band labelled by i, associated with the pixel in column q of row p of the image. Assuming that the image is a square N by N image, so that $1 \le p \le N$ and $1 \le q \le N$, and that there are n bands, so that $1 \le i \le n$, image data in the form of two-dimensional arrays take the place of population parameters. The complete range of subscripts $1 \le p \le N$ and $1 \le q \le N$ now corresponds to the population and each band image corresponds to one of the parameters measured for the population. A set of n intensity values now exists corresponding to the n bands of the multi-spectral scanner, for each value of the pair of subscripts p and q. A particular pair (p,q) in this case is the analogue of one member of the population in the original psychological formulation of Hotelling.

Each band of the image can be thought of as a one-dimensional array or vector, x_i or $x_i(j)$ where $1 \le j \le N^2$ instead of thinking of each band of the image as a two-dimensional array, $I_i(p,q)$. For the sake of argument it may be supposed that the first N components of x_i are constructed from the first row of $I_i(p,q)$, the second N components from the second row and so on. Thus

$$x_i = \{I_i(1,1), I_i(1,2),...I_i(1,N),...$$

$$...I_i(N,1), I_i(N,2),...I_i(N,N)\} \qquad (9\text{-}16)$$

All the image data are now contained in this set of N vectors x_i, where each vector x_i is of dimension N^2.

The covariance matrix C_x of the vectors x_i is now defined as

$$C_x = E\{(x_i - \bar{x}_i)(x_i - x_i')\} \qquad (9\text{-}17)$$

where $\bar{x}_i = E(x_i)$, the expectation, or mean, of the vectors x_i and the prime used to denote the transpose. From the data the mean and the covariance can be estimated:

$$\bar{x}_i = \frac{1}{n} \sum_{i=1}^{n} x_i \qquad (9\text{-}18)$$

and

$$C_x = \frac{1}{n} \sum_{i=1}^{n} (x_i - \bar{x}_i)(x_i - \bar{x}_i)'$$

$$= \frac{1}{n} \left\{ \sum_{i=1}^{n} x_i x_i' \right\} - \bar{x}_i \bar{x}_i' \qquad (9\text{-}19)$$

The mean vector will be of dimension N^2 and the covariance matrix will be of dimension $N^2 \times N^2$.

The objective of the Hotelling transformation is to diagonalize the covariance matrix, that is to transform from a set of bands that are highly correlated with one another to a set of uncorrelated bands or principal components. In order to achieve the required diagonalization of the covariance matrix a transformation is performed using a matrix A where the elements of A are the components of the normalized eigenvectors of C_x. That is

$$A = \begin{bmatrix} e_{11} & e_{12} & e_{13} & \cdots & e_{1N^2} \\ e_{21} & e_{22} & e_{23} & \cdots & e_{2N^2} \\ \vdots & & & & \\ e_{N^2 1} & e_{N^2 2} & e_{N^2 3} & \cdots & e_{N^2 N^2} \end{bmatrix} \qquad (9\text{-}20)$$

where e_{ij} is the jth component of the ith eigenvector. The Hotelling transformation then consists of replacing the original vector x_i by a new vector y_j where

$$y_i = A(x_j - \bar{x}_j) \qquad (9\text{-}21)$$

and the transformed covariance matrix C_y, which is now diagonal, is related to C_x by

$$C_y = A C_x A' \qquad (9\text{-}22)$$

where A' denotes the transpose of A. The example of a multi-spectral image containing areas corresponding to water, vegetation-covered land and built-over land might be used to indicate what is involved in a slightly more concrete fashion. To distinguish among these three categories an attempt could be made to identify each area in the data from a single band. An attempt could also be made to carry out a multi-spectral classification, in which case there will be a need for some evidence external to the digital data of the image itself in order to identify the classes. By using the principal components transformation the maximum discrimination between different classes can be achieved without any reference to external evidence outside the data set of the image data itself.

9.9 *Fourier transforms*

The use of Fourier transforms for the removal of noise from images is an accepted method of image processing.

To establish the notation,

$$F(k) = \int_{-\infty}^{\infty} f(x) \, e^{2\pi i k x} \, dx \qquad (9\text{-}23)$$

is written for a function $f(x)$ of one variable x and

$$F(k_x, k_y) = \int_{-\infty}^{\infty} \int_{-\infty}^{\infty} f(x,y) \, e^{2\pi i (k_x x + k_y y)} \, dx \, dy \qquad (9\text{-}24)$$

for a function $f(x,y)$ of two variables x and y.

If an image is held on a film transparency the Fourier transform may be obtained optically. If the image has been digitized and is represented by a function $f(x,y)$ of two variables then the Fourier transform has to be calculated numerically. Irrespective of the actual method used for performing the Fourier transform, the original function can be reconstructed by performing the inverse Fourier transform. For a function $f(x,y)$ of two variables which represents a perfect image the process of taking the Fourier transform and then doing a second transformation on this transform to regenerate the original image is bound to lead to a degeneration of the quality of the image. If optical processing is used the degradation will arise from aberrations in the optical system and if digital processing is used the degradation will arise from rounding errors in the computer and from truncation errors in the algorithms. It may, therefore, seem strange that the quality of an image might be enhanced by taking a Fourier transform of the image and then taking the inverse Fourier transform of that transform to regenerate the image again. However, the basic idea is that it may be easier to identify spurious effects in the Fourier transform than in the original image. These spurious effects can then be removed, a process referred to as filtering. Having filtered the transform to remove the imperfections, the image can then be reconstructed by performing the inverse Fourier transform. An improvement in the quality of the image is then often obtained in spite of the aberrations or numerical errors.

An image may simply be lacking in contrast, it may contain random noise or it may contain a periodic blemish such as striping. A commonly-encountered example of striping is found in LANDSAT multi-spectral scanner (MSS) images. The LANDSAT MSS is constructed in such a way that six scan lines of the image are generated simultaneously using an array of six different detectors for each spectral band, 24 detectors in all. Although the six detectors for any one band are nominally identical they are inevitably not exactly so and consequently there may be a striping with a periodicity of 6 lines in the image, (see Figure 9-16).

In this example the periodic feature is a blemish requiring removal. But, of course, there are other situations in which the Fourier transform of an image may be studied in order to try to identify a periodic feature that is not very apparent from an inspection of the original image itself. This is very much what is done in X-ray and electron diffraction work in which the diffraction pattern is used to identify or quantify the periodic structure of the material that is being investigated. Similarly, Fourier transforms of images of wave patterns on the surface of the sea, obtained from aerial photography or from satellites, are used to find the wavelengths of the dominant waves present.

The Fourier transform for a very simple function of one variable

$$f(x) = 0 \quad -\infty < x < -a \quad \text{and} \quad a < x < \infty$$

$$f(x) = 1 \quad -a < x < a \tag{9-25}$$

is shown in Figure 9-17; this will be recognized in optical terms as corresponding to the intensity distribution in the diffraction pattern from a

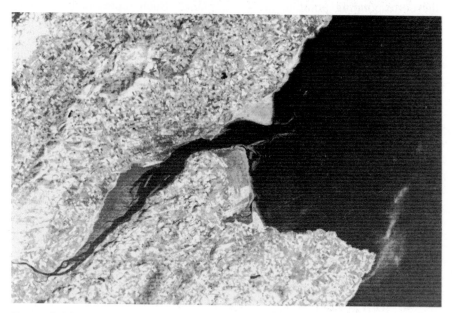

Figure 9-16 Example of a LANDSAT MSS image showing the 6-line striping effect, band 5, path 220, row 21, of 24 October 1976, of the River Tay, Scotland (Cracknell et al., 1982)

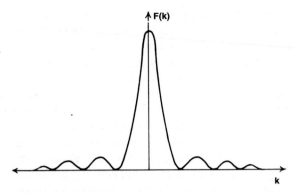

Figure 9-17 Standard diffraction pattern obtained from a single slit aperture, k=*spatial frequency,* F(k)=*radiance*

single slit. For a function of two variables the corresponding Fourier transform $F(k_x,k_y)$ will be an analogous function of the two variables k_x and k_y, see Figure 9-18.

As it is rather inconvenient to deal with a representation of the Fourier transform as a function of two variables using three dimensions in space, it is more common to represent the Fourier transform as a grey scale image in which the value of the transform $F(k_x,k_y)$ is represented by the intensity at the corresponding point in the k_x, k_y plane. Such representations of the Fourier transform are very familiar to physicists who encounter them frequently as films of optical, X-ray or electron diffraction patterns. Strictly speaking the Fourier transform is a complex function having modulus and

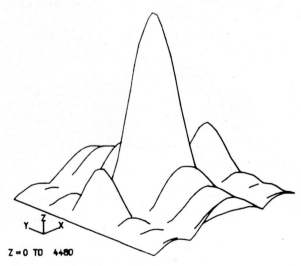

Figure 9-18 Example of two-dimensional Fourier transform

argument (or amplitude and phase); in the photographic representation the phase only is recorded and not the amplitude information at each point in the transform space.

It is, perhaps, appropriate to take a brief account of the optical process of taking a Fourier transform. In taking a Fourier transform of a two-dimensional object, such as a film image of some remotely sensed scene, one is essentially analysing the image into its component spatial frequencies. This is what a converging lens does when an object is illuminated with a plane parallel beam of coherent light. The complex field of amplitude and phase distribution in the back focal plane is the Fourier transform of the field across the object; in observing the diffraction pattern or in photographing it, one is observing or recording the intensity data and not the phase data. Actually, to be precise, the Fourier transform relation is only exact when the object is situated in the front focal plane; for other object positions phase differences are introduced, although these do not affect the appearance of the diffraction pattern. It will be clear that, since rays of light are reversible, the object is the inverse Fourier transform of the image. The inverse transform can thus be produced physically by using a second lens. As already mentioned, the final image produced would, in principle, be identical to the original object although it will actually be degraded as a result of the aberrations in the optical system. This arrangement has the advantage that by inserting a filter in the plane of the transform the effect of that filter on the reconstructed image can be seen directly and visually, see Figure 9-19.

The effects that different types of filters have when the image is reconstructed can thus be studied quickly and effectively. This provides an example of a situation in which it is possible to investigate and demonstrate effects and principles much more simply and effectively with optical image processing techniques than with digital methods.

The effect of some simple filters can be illustrated with a few examples

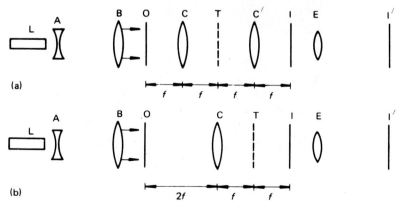

Figure 9-19 Two optical systems suitable for optical filtering (Wilson, 1981)

that have been obtained by optical methods. A spatial filter is a mask or transparency that is placed in the plane of the Fourier transform, i.e. at T in Figure 9-19, and various types of filter can be distinguished:

- A blocking filter (i.e. a filter that is simply opaque over part of its area);

- An amplitude filter;

- A phase filter;

- A real-valued filter, that is a combination of an amplitude filter and a phase filter, where the phase change is either 0 or π; and

- A complex-valued filter which can change both the amplitude and the phase.

A blocking filter is, by far, the easiest type of filter to produce. Figure 9-20(a) shows an image of an electron microscope grid and Figure 9-20(b) shows its (optical) Fourier transform. Figure 9-20(c) shows the transform with all the non-zero k_y components removed. Consequently, when the inverse transform is taken no structure remains in the y direction, see Figure

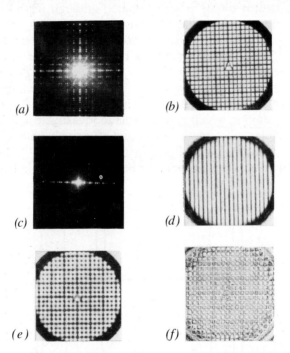

Figure 9-20 (a) The optical transform from an electron microscope grid; (b) image of the grid; (c) filtered transform; (d) image due to (c); (e) image due to zero-order component and surrounding four orders; and (f) image when zero-order component is removed by blocking (Wilson, 1981)

9-20(d). The effects of two other blocking filters are shown in Figure 9-20(e) and (f). The six-line striping present in LANDSAT MSS images has already been mentioned. By using a blocking filter to remove the component in the transform corresponding to this striping one can produce a de-striped image. The removal of a raster from a televison picture is similar to this, see Figure 9-21. There is also the possibility of removing the result of screening a half-tone image; the diffraction pattern from a half-tone object will contain a two-dimensional arrangement of discrete maxima, with the transform of the picture centred on each maximum. A filter which blocks out all except one order will produce an image without the half-tone. This approach can also be applied to the smoothing of images that have been produced on computer output devices such as lineprinters, teletypes and dot-matrix printers, (see Figure 9-22).

The more high spatial frequencies there are present in the Fourier transform, the more fine detail can be accounted for in an image. A filter which allows only the low spatial frequencies to pass and to contribute to the reconstructed image is called a low-pass filter. A filter which allows

Figure 9-21 Raster removal (Wilson, 1981)

Figure 9-22 Half-tone removal (Wilson, 1981)

only the high spatial frequencies to pass is called a high-pass filter. A high-pass filter leads to edge enhancement of the original image because it is the high spatial frequencies which are responsible for the sharp edges, (see Figure 9-23).

If an image contains random noise, rather than periodic blemishes such as striping, a raster, or half-tone screening, it is still possible to improve the image using Fourier transforms. This is illustrated by a hypothetical noise-free image that gives rise to the Fourier transform shown in Figure 9-24. In this transform the central peak is very much larger than any other peak and the size of the peaks decreases as one moves further away from the centre. If some random noise is now introduced into the image and the transform taken of the noisy image, a transform such as that shown in Figure 9-25 will be obtained. Now the size of the off-centre peaks has increased relative to the size of the central peak and the peaks do not necessarily become smaller

Figure 9-23 Edge enhancement by high-pass filtering (Wilson, 1981)

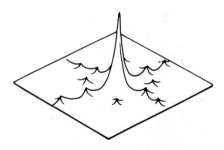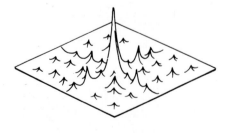

Figure 9-24 Fourier transform for a hypothetical noise-free image

Figure 9-25 Fourier transform shown in Figure 9-24 with the addition of some random noise

as one moves further away from the origin. A suitable filter to apply to this transform to eliminate the noise from the regenerated image would be a blocking filter which allows large discrete maxima to pass but blocks the small peaks.

Consider, briefly, the forms of some filters from the point of view of digital rather than optical processing. Here it is supposed that the process of filtering consists of evaluating the product $H(k_x, k_y) \times F(k_x, k_y)$, where the first factor is the filter function and the second is the Fourier transform. Four common types of filter function are illustrated in Figure 9-26 where

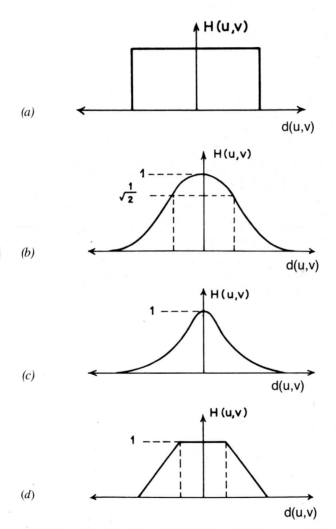

Figure 9-26 Four filter functions (a) ideal filter; (b) Butterworth filter; (c) exponential filter; and (d) trapezoidal filter

the functions are radially symmetric, that is they are simply functions of d where

$$d = \sqrt{\frac{u^2 + v^2}{k_x^2 + k_y^2}} \qquad (9\text{-}26)$$

These are relatively simple functions; more complicated filters can be used for particular purposes.

10 Applications of remotely sensed data

10.1 Introduction

Remotely sensed data can be used for a great variety of practical applications, all of which relate, in general, to Earth resources. For convenience, and because the innumerable applications are so varied and far reaching, in this chapter these applications are classed into seven major categories, each coming under the purview of some recognized professional discipline or speciality. These categories include applications to

- The atmosphere;
- The geosphere;
- The biosphere;
- The hydrosphere; and
- The cryosphere; together with
- Environmental applications; and
- Applications of data collection systems.

This arrangement is not entirely satisfactory as some disciplines such as cartographic mapping have their own sets of unique applications, but also rely upon observations and measurements that overlap with, and are of mutual interest to, other disciplines. Since many examples of applications of remotely sensed data exist, their treatment here can only be of a cursory nature. The set of applications that is described in this chapter is by no means exhaustive. Many additional examples exist, both outwith and within the categories mentioned, see also Table 1-2.

10.2 Applications to the atmosphere

10.2.1 Weather satellites in forecasting and nowcasting

Weather forecasters need access to information from large areas as quickly and as often as possible because weather observations rapidly become out of date. Satellites are best able to provide the kinds of data that satisfy these requirements in terms of both coverage and immediacy. A good description of the current weather situation is essential to successful short-period weather forecasting, particularly for forecasting the movement and development of precipitation in the next six hours. Satellite pictorial data are particularly useful in that they provide precision and detail for short-period weather forecasting. The data allow synoptic observations to be made of the state of the atmosphere, from which a predicted state may be interpolated on the basis of physical understanding and past experience of the way in which the atmosphere behaves.

Meteorologists have been making increasing use of weather satellite data as aids for analysing synoptic and smaller-scale weather systems since 1960. The use and importance of satellite data has increased

- With the continued improvement of satellite instrumentation; and

- Because of the extra dependence placed on them following the reduction in the number of ocean weather stations making surface and upper-air observations.

Indeed, in regions where more conventional types of surface and upper-air observations are few, or lacking (i.e. oceanic areas away from the main airline routes, interiors of sparsely populated continents, etc.), satellite data at times provide the only current or recent evidence pertaining to a particular weather system.

Satellite observations are now regularly used in weather forecasting and nowcasting, alongside observations made from land stations, ships, aircraft and by balloon-borne instruments. Commonly, weather satellites produce visible and infrared images. These are the pictures normally associated with television presentations of the weather. The relative importance of the satellite observations depends on the weather situation. A skilled forecaster has a very good understanding of the relationship between the patterns in maps of temperature, pressure and humidity and the location (or absence) of active weather systems. Although there is not a unique relationship between a particular cloud system and the distribution of the prime variables, the relationships are fairly well defined and confined within certain limits. This means that the forecaster can modify the analyses maps to be consistent with the cloud systems as revealed by the satellite data.

'Nowcasting' is the real-time synthesis, analysis, and warning of significant — chiefly hazardous — local and regional weather based on a combination of observations from satellites, ground-based radar and dense ground networks reporting through satellites. The trend towards nowcasting, enabled by remote sensing technologies, is developing as a response to the need for timely information in disaster avoidance and management and for numerical models of the atmosphere. The improvement of flash-flood and tornado warnings, or the monitoring of the dispersal of an accidental radioactive release, illustrate the call on immediate weather information.

10.2.2 Weather radars in forecasting

Weather radars make it possible to track even small-scale weather patterns and individual thunderstorms. However, the range limitations of radar and the curvature of the Earth mean that a single ground-based weather radar can observe a travelling rainfall system for only a limited period, and then often only part of that system. In the U.K. a network of weather radars has been set up to overcome these limitations, (see Figure 10-1). This network is used to provide up-to-date information on the distribution of surface rainfall at intervals of 15 minutes. Detailed forecasts for specific locations may be attempted by replaying recent radar image sequences to reveal the movement of areas of rain which leads naturally to the prediction of their future movement.

Satellite data may be used to extend the radar coverage. The U.K.

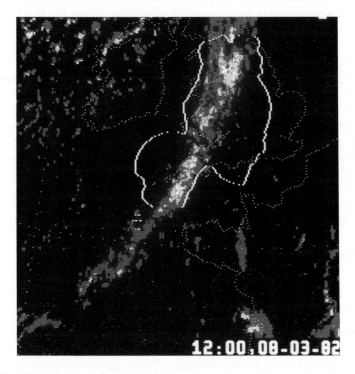

Figure 10-1 Distribution of rainfall during the passage of a cold front across the British Isles as inferred from weather radars (within the white outline) and METEOSAT (elsewhere) (UK Meteorological Office)

Meteorological Office's weather radar network uses half hourly METEOSAT imagery to identify rain clouds outside the radar coverage area for inclusion in the forecast. The radar and satellite pictures are registered to the same map projection which enables the integration of the two types of remotely sensed weather data. The relationship between the rain at the surface, as identified by the radar, and the cloud above, as identified in the METEOSAT imagery, can then be examined by forecasters. Because the correspondence between cloud and rain is variable, the radar data are used to 'calibrate' the satellite images in terms of rainfall. Since rainfall patterns can be inferred in only a very broad sense from satellite data, the radar data are used in preference to the satellite data where both are available.

One of the more widely recognized problems in radar meteorology is that the variability of the drop size distribution causes the relationship between returned signal and rainfall intensity to vary. This is because for the wavelengths commonly used in weather radars (about 5 cm) raindrops behave as Rayleigh scatterers and their reflectivity depends on the sixth power of the drop diameter. However, as far as the forecaster is concerned, the value of radar lies not so much in its ability or otherwise to measure rainfall accurately at a point as in its ability to define the field of surface rainfall semi-quantitatively over extended areas. One important category of error is caused by the radar beam observing precipitation some distance above the ground, especially at long ranges. Thus the radar measurements may either underestimate or overestimate the surface rainfall according to whether the precipitation accretes or evaporates, or indeed changes from snow to rain, below the radar beam. Since these errors are due to variations in the physical nature of the phenomenon no purely objective technique can be used to correct them on all occasions, but they can to some extent be corrected subjectively in the light of other meteorological information. The error of radar-derived rainfall totals depends on the density of the rain gauge network used in comparison. As a consequence, there is less error when widespread and uniform rain is compared than isolated showers where there may be poor agreement because of the sparcity of rain gauges. Indeed, an isolated downpour may not even be recorded by conventional methods.

10.2.3 The determination of temperature changes with height from satellites

In pictorial form, weather satellite data is capable of revealing excellent resolution in the position, extent and intensity of cloud systems. Pictorial data, however, have to be interpreted by experienced forecasters and parameterized to enable the mapping of the quasi-horizontal fields of pressure, wind, temperature and humidity at several discrete levels in the troposphere. Since satellites orbit far above the region of the atmosphere which contains the weather, techniques for obtaining information about the atmosphere itself are limited by the fact that the observations that are most needed are not measured directly. As a result it is, for the moment, not possible to obtain the vertical resolution which is desirable in temperature profiles or winds.

As well as showing the size, location and shape of areas of cloud, visible and infrared satellite pictures can, from an examination of the relative brightness and texture of the images, also provide information on the vertical structure of clouds. Brightness of a cloud image on a visible picture depends upon the Sun's illumination, the reflectivity (related to cloud thickness) and the relative positions of cloud, Sun and radiometer. On the infrared picture the brightness depends on the temperature of the emitting surface with the brighter the image, the colder (higher) the cloud top. Infrared imagery is obtained in regions of the electromagnetic spectrum where the atmosphere is nearly transparent, so that radiation from the clouds and surface should reach the satellite relatively unaffected by the intervening atmosphere. However, the vertical distribution of atmospheric temperature may be inferred by measuring radiation in spectral regions where the atmosphere is absorbing. If the vertical temperature distribution is known, the distribution of water vapour may be inferred. These measurements, however, are technically difficult to make. The variation of temperature is obtained by making measurements of emission at several wavelengths with different strength of atmospheric absorption.

Temperature profiles, expressed as mean temperatures through layers of depth 1.5 – 2 km, are produced from the data from the TOVS system (see Section 8.4.1), at 250 km grid spacing, with provision for 500 km grid spacing for reduced coverage. Moisture soundings are also produced for these layers. Comparisons with radiosonde observations reveals differences of about 2° Celsius (rms). The limiting factors are

- The raw observations relate to deep layers (defined by the atmospheric density profile);

- The influence of clouds;

- Global coverage is built up over several hours;

- It is particularly difficult to obtain temperatures and winds in and around frontal zones, where they are of most importance; and

- The error characteristics of satellite observations are very different from those of conventional observations, necessitating different analysis procedures for the best use to be made of the data.

Even with these limitations satellites are helping meteorologists both by providing a considerably increased total number of observations than could have been obtained by conventional means and also by providing these observations at increased levels of consistency.

10.2.4 Measurements of windspeed

10.2.4.1 Tropospheric estimations from cloud motion

Some clouds move with the wind. If these clouds can be seen and their geographical position determined in two successive satellite pictures then the displacement can be used to determine the speed of the wind at the level of the cloud. This simple principle forms the basis for the derivation of several thousand wind observations each day from the set of weather satellites.

Small clouds are most likely to move with the wind, but these are generally too small to be detected by the radiometers on satellites. Moreover, their life-cycle can be shorter than the half-hourly interval between successive images recorded by a geostationary weather satellite. Consequently larger clouds, and more often, patterns of cloud distribution 10 – 100 km in size are used. The longer the time interval between the pair of images, the greater the displacement and (up to a point), the more accurate the technique becomes. Gauged against radiosonde wind measurements, satellite winds derived by present techniques have an accuracy of 3 ms^{-1} at low levels. However, at upper levels the scatter in differences of wind velocities as measured by satellite and sonde is substantially larger. It is obviously necessary to determine the height to which the satellite-derived wind should be assigned. The only information available from the satellite for this purpose are measurements of radiation relating to the cloud top. From these the temperature of the cloud top may be deduced and, given a knowledge of the temperature variation with height, the cloud-top height may be derived. This process is subject to error, especially if the cloud is not entirely opaque which, unfortunately, is often the case for cirrus ice-clouds in the upper troposphere. Accordingly, the transmissivity of cirrus clouds needs to be estimated in order to obtain the proper cloud top heights. An empirical correction is sometimes used, but METEOSAT has a water vapour channel which helps to reduce uncertainties.

10.2.4.2 Microwave estimations of surface wind shear

The best measurements of surface wind velocity from satellites are made by radars which observe the scatter of centimetre-wavelength radio waves from small capillary waves on the sea surface. Wind speed is closely related to the flux of momentum to the sea. Accordingly, the amount of scatter per unit area of surface, the scattering cross section, is highly correlated with wind speed and direction at the surface. Scatterometers were operated for a few hours on SKYLAB and for three months on SEASAT in 1978. The SEASAT Scatterometer System (SASS), produced excellent maps of surface wind with a resolution of 50 km and an accuracy of ± 2 ms^{-1} in speed and $\pm 20°$ in direction over the range of 4 to 24 ms^{-1} in windspeed.

While the altimeters flown on SKYLAB, GEOS-3, and SEASAT were primarily designed to measure the height of the spacecraft above the Earth's surface, the surface wind speed, although not direction, can be inferred from the shape of the returned altimeter pulse. Similarly, the Scanning Multi-channel Microwave Radiometer (SMMR) flown on SEASAT and NIMBUS-7 was primarily intended to measure sea-surface temperature, but surface wind speed is extracted using an algorithm based on the measured response in all channels of the instrument. Again, no indication of direction is provided.

Space-borne radar observations of wind will be resumed at the end of the decade when Canada's RADARSAT, the European Space Agency's ERS-1 and the Japanese ERS-1 satellites become operational.

10.2.4.3 Sky wave radar

An attempt to evaluate the use of sky wave radar techniques for the determination of wind and sea-state parameters (see Chapter 3) was provided by the Joint Air-Sea Interaction (JASIN) project which was carried out in the summer of 1978, see Shearman (1981). Sky wave observations of the JASIN area in the North Atlantic, see Figure 10-2, were made by Birmingham University and the Rutherford Appleton Laboratory. Simultaneous *in situ* data were collected by fourteen oceanographic vessels in the area. The observations were all made during daylight hours and the radio waves were reflected once by the sporadic E layer at a height of about 100 km. Initially, earlier published relationships (due to Maresca and Barnum, 1977) between the width of the larger Bragg line at -10 dB and wind speed were applied to the data (Shearman *et al.*, 1979). This gave good agreement between the results of sky wave radar determinations of wind direction and ship-made measurements for wind speeds of about 10 ms^{-1}, but poor results were obtained when the same relationships were used for data corresponding to lower windspeeds.

10.2.5 Hurricanes

Satellite observations, together with land-based radar, are used extensively in forecasting severe weather. Major thunderstorms, which may give rise to tornadoes and flash-floods, can often be identified at a stage where warnings can be issued early enough to reduce loss of life and property. In more remote ocean areas, satellite observations can provide early location of hurricanes (tropical storms) and enable their development

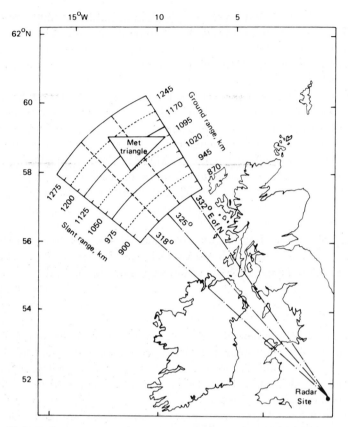

Figure 10-2 Radar coverage during the JASIN experiment (Shearman, 1981)

and movement to be monitored. In the last decade or two, probably no hurricane or tropical storm anywhere in the world has gone unnoticed by weather satellites, during which time much has been learned about the structure and movements of these small but powerful vortices from the satellite evidence. Satellite-viewed hurricane cloud patterns enable the compilation of very detailed classifications and the determination of maximum wind speeds. The use of enhanced infrared imagery in tropical cyclone analysis adds objectivity and simplicity to the task of determining tropical storm intensity. Satellite observations of tropical cyclones are used to estimate their potential for spawning hurricanes. The infrared data not only afford continuous day/night storm surveillance but also provide quantitative information about cloud features that relate to storm intensity; thus, cloud-top temperature measurements and temperature gradients can be used in place of qualitative classification techniques employing visible wavebands. The past history of cloud pattern evolution and knowledge of the current intensity of tropical storms is very useful for predicting their developmental trend over the next 24 hours and allow an early warning capability for shipping and areas over land in the path of potential hurricanes.

Hurricanes and typhoons exhibit a great variety of cloud patterns, but most can be described as having a comma configuration. The comma tail is composed of convective clouds that appear to curve cyclonically into a centre. As the storm develops, the clouds are observed to form bands that wrap around the storm centre producing a circular cloud system that usually has a cloud-free, dark eye in its mature stage (Figure 10-3). The intensity of hurricanes is quantifiable either by measuring the amount by which cold clouds circle the centre or by using surrounding temperature and eye criteria. Large changes in cloud features are related to the intensity while increased

Figure 10-3 STS41-C onboard scene of Earth. Hurricane Kamysi in the Indian Ocean, northeast of Madagascar (NASA)

encirclement of the cloud system centre by cold clouds is associated with a decrease in pressure and increase in windspeed.

Weather satellites have almost certainly justified their expense through the assistance they have given in hurricane forecasting alone. The damage caused by a single hurricane in the U.S.A. is often of the order of billions of dollars. In 1983, Hurricane Alicia caused an estimated $2.5 billion of damage and was responsible for 1804 reported deaths and injuries. In November 1970 a tropical cyclone struck the head of the Bay of Bengal and the loss of life caused by the associated wind, rain and tidal flooding exceeded 300,000 people. Indirectly, this was the trigger which led to the establishment of an independent state of Bangladesh. Clearly, timely information about the behaviour of such significant storms may be almost priceless.

10.2.6 Satellite climatology

The development of climatology as a field of study has been hampered by the inadequacy of available data. Satellites are helping enormously to correct this deficiency as they afford more comprehensive and more dynamic views of global climatology than were possible before. In pre-satellite days certain components of the radiation balance, such as short wave (reflected) and long wave (absorbed and re-radiated) energy losses to space, were established by estimation, not measurement. The only comprehensive maps of global cloudiness compiled in pre-satellite days depended heavily on indirect evidence and could not be time-specific. While satellite-derived climatological products have only been available for two decades and are accordingly limited in their use for longer term trend analysis, these products are becoming increasingly interesting and valuable as the databases are built up. These databases include inventories of parameters used in the determination of:

- The earth/atmosphere energy and radiation budgets, particularly the net radiation balance at the top of the atmosphere which is the primary driving force of the Earth's atmospheric circulation;

- Global moisture distributions in the atmosphere, which relates to the distribution of energy in the atmosphere;

- Global temperature distributions over land and sea, and accordingly the absorption and radiation of heat;

- The distribution of cloud cover, which is a major influence on the albedo of the earth/atmosphere system and its component parts, and an indicator of horizontal transport patterns of latent heat;

• Sea-surface temperatures, which relate directly to the release of latent heat through evaporation;

• Wind flow and air circulations, which relate to energy transfer within the earth/atmosphere system;

• The climatology of synoptic weather systems, including their frequencies and spatial distribution over extended periods.

The World Climate Research Programme (WCRP) aims to discover how far it is possible to predict natural climate variation and man's influence on the climate. Satellites contribute by providing observations of the atmosphere, the land surface, the cryosphere and the oceans with the advantages of global coverage, accuracy and consistency. Quantitative climate models will enable the prediction and detection of climate change in response to pollution and resolve the speculation about the "greenhouse" effect. In addition to the familiar meteorological satellites, novel meteorological missions are now established to support the Earth Radiation Budget Experiment (ERBE) and the International Satellite Cloud Climatology Project (ISCCP).

Figures 10-4 (a) and (b) show monthly mean cloud amount at local midnight derived from the 11.5 μm channel of the Temperature Humidity Infrared Radiometer on NIMBUS-7. Figure 10-4(a), for January 1980, represents normal conditions and Figure 10-4(b), for January 1983, represents conditions in an El Nino year. Analysis showed that during the period from December 1982 to January 1983, the equatorial zonal mean cloud amount is 10% higher than in a non-El Nino year. The most significant increases occurred in the eastern Pacific Ocean.

During the 1982 – 1983 El Nino event, significant perturbations in a diverse set of geophysical fields occurred. Of special interest are the planetary-scale fields that act to modify the Outgoing Longwave Radiation (OLR) field at the top of the atmosphere. The most important is the effective "cloudiness": specifically, perturbations from the climatological mean of cloud cover, height, thickness, water content, drop/crystal size distribution and emissivity. Also important are changes in surface temperature and atmospheric water vapour content and, to a lesser extent, atmospheric temperature. Changes in one or more of these parameters, regionally or globally, cause corresponding anomalies in the broad-band OLR at the top of the atmosphere.

To facilitate the examination of the time evolution of the El Nino event from the perspective of the set of top of the atmosphere OLR fluxes, monthly-averaged time-anomaly fields have been generated from observations derived from the NIMBUS-7 ERB. These are defined in terms of departures (Wm^{-2}) from the climatology for that month. The term "climatology" is used somewhat loosely here to indicate the two years between June 1980 and May 1982. Thus, a two-year mean pre-El Nino

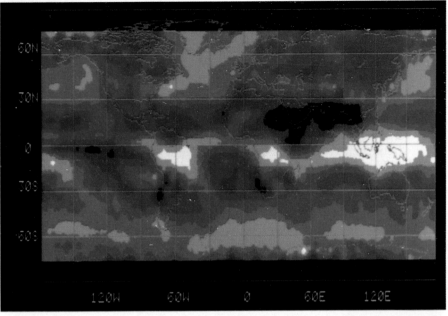

(a)

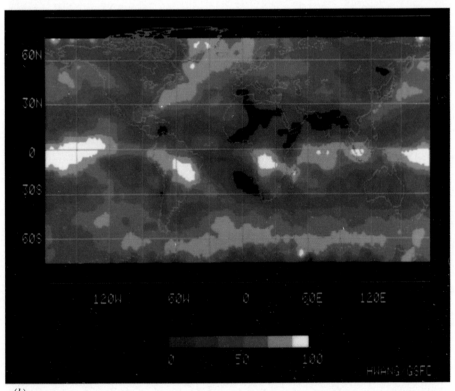

(b)

Figure 10-4 Monthly mean cloud amount at midnight in January derived from
NIMBUS-7 Temperature Humidity Infrared Radiometer's (THIR's) 11·5 μm channel
data; (a, upper) in a normal year (1980) and (b, lower) in an El Nino year (1983)
(NASA Goddard Space Flight Center)

seasonal cycle is removed in the creation of the anomaly maps. Analysis of these fields indicates that the OLR anomaly response to the El Nino event of 1982-1983 can be characterized as having four modes of behaviour: onset, intensification, expansion and decay.

In January the peak amplitudes of the OLR anomalies are generally reached. The negative radiation centre in the equatorial Pacific has reached -88 Wm^{-2}. To its north and south, the accompanying positive anomalies now average half its magnitude. An interesting large-amplitude pattern exists along the equator. The three areas that are normally quite active, convectively, at this time of the year are Indonesia, the Amazon river basin and the Congo river basin. They now show positive OLR anomalies indicative of reduced convection. These are replaced, instead, with the negative anomalies over the Arabian Sea, the Indian Ocean and the central equatorial Pacific Ocean. The centre over Europe has intensified, while the centre over the United States has now moved into the Gulf of Mexico. At this time the true global nature of the El Nino event is evident.

Figure 10-5 (a) and (b) are the first global maps ever made of the Earth's mean skin temperature for day and night. The images were obtained by a team of NASA scientists from the Jet Propulsion Laboratory in Pasadena, California, and the Goddard Space Flight Center in Greenbelt, Maryland. The satellite data were acquired by the High Resolution Infrared Sounder and the Microwave Sounding Unit, both instruments flying on board the NOAA weather satellites. The surface temperature was derived from the 3.7 μm window channels in combination with additional microwave and infrared data from the two sounders. The combined data sets were computer processed using a data analysis method that removed the effects of clouds, the atmosphere and the reflection of solar radiation.

The ocean and land temperature values have been averaged spatially over a grid of 2°30′ latitude by 3° longitude and correspond to the month of January 1979. The mean temperature values for this month clearly show several cold regions, such as Siberia and northern Canada, during the northern hemisphere's winter, and a hot Australian continent during the southern hemisphere's summer. Mountainous areas are clearly visible in Asia, Africa and South America. The horizontal gradients of surface temperature displayed on the map in colour contour intervals of 2°C show some of the major features of ocean-surface temperature, such as the Gulf Stream, the Kuroshio Current, and the local temperature minimum in the eastern tropical Pacific Ocean. The satellite-derived sea-surface temperatures are in very good agreement with ship and buoy measurements.

Surface temperature data are important in weather prediction and climate studies. Since the cold polar regions cover a small area of the globe relative to the warm equatorial regions the mean surface temperature is dominated by its value in the tropics. The mean calculated skin-surface temperature during January 1979 is given in Table 10-1.

Climatologists are testing the accuracy of using surface temperature

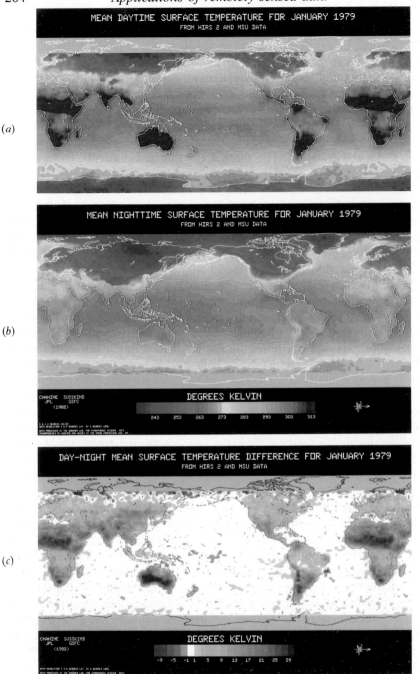

Figure 10-5 Mean day and night surface temperatures derived from sounder data (image provided by Jet Propulsion Laboratory): (a, top) daytime temperature; (b, centre) nighttime temperature; and (c, bottom) mean temperature difference

*Table 10.1 Mean skin surface temperature during
January 1979*

Area	temperature (°C)
Global	14·14
N. hemisphere	11·94
S. hemisphere	16·35

anomalies as potential predictors, on seasonal time scales, of weather conditions over different parts of the world.

Figure 10-5(c) shows the monthly average of the differences between day and night temperature. This difference map provides striking contrast between oceans and continents. The white area indicates day – night temperature differences in the range ±1 K. This small difference indicates ocean areas, having high heat capacity and a large degree of homogeneity, while areas with larger day – night temperature differences are continental landmasses. The outlines of all continents can be plotted accordingly on the basis of the magnitude of the difference between day and night temperature. The day – night temperature differences over land clearly distinguish between arid and vegetated areas and may indicate soil moisture anomalies.

Figure 10-6 (a) and (b) show monthly mean total atmospheric water vapour derived from observations by the NIMBUS-7 Scanning Multi-channel Microwave Radiometer (SMMR). January 1980, Figure 10-6(a), represents a normal year while January 1983, Figure 10-6(b), is an El Nino year. Time-longitude cross section analysis showed that during the period from December 1982 to January 1983, the water vapour content increased by more than 1.2 g/cm^2 compared to the same month of the normal year in the latitude band of ±5°. The most significant increases occurred at 100° – 160°W in the Pacific Ocean.

The series of pictures shown in Figure 10-7 has been produced using data from the Total Ozone Mapping Spectrometer (TOMS) instrument on the NIMBUS-7 satellite. These pictures show progressive development of an ozone hole in the month of October in the Antarctic. The ozone column amount in the centre of the hole has decreased by more than 50% in less than five years. The precise cause for the development of this hole is unknown, but there are speculations that the hole is created by some poorly understood heterogeneous chemical reaction that takes place in the presence of photochemically active pollutants such as chlorofluoromethanes trapped in a strong polar vortex. The disappearance of the ozone hole early in November coincides well with the break up of the polar vortex.

The chief advantages for climatology of satellite remote sensing systems are:

- Weather satellite data are far more complete for the whole globe than conventional data;

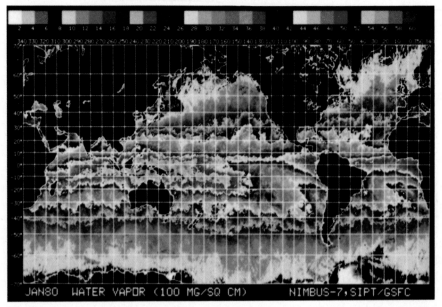

(a)

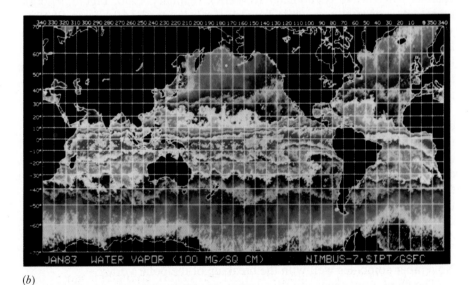

(b)

Figure 10-6 Monthly mean total atmospheric water vapour derived from NIMBUS-7 Scanning Multi-channel Microwave Radiometer (SMMR) data in January (a) in a normal year (1980) and (b) in an El Nino year (1983)

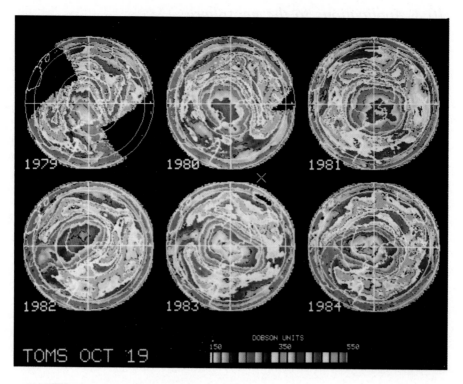

Figure 10-7 Progressive development of an ozone hole in the month of October over the Antarctic continent derived from NIMBUS-7 Total Ozone Mapping Spectrometer (TOMS) data (NASA Goddard Space Flight Center)

- Satellite data are more homogeneous than those collected from a much larger number of surface observatories;

- Satellite data are often spatially continuous, as opposed to point recordings from the network of surface stations;

- Satellites can provide more frequent observations of some parameters in certain regions, especially over oceans and high latitudes;

- The data from satellites are collected objectively, unlike some conventional observations e.g. visibility and cloud cover;

- Satellite data are immediately amenable to computer processing.

10.3 Applications to the geosphere

10.3.1 Geological information from electromagnetic radiation

In the electromagnetic spectrum, the remote sensor acquiring the shortest wavelength data is the gamma-ray spectrometer. The gamma-ray spectrometer senses the radiometric environment and can acquire data on soil composition where conditions such as moisture content are known, or conversely, moisture content where the soil composition is known. In modern systems, digital data are collected which can be used to provide a computer-map of the soil and rock exposures environment of an area through the covering vegetation in terms of relative percentages of uranium, potassium and thorium, (see Figure 10-8).

Remote sensors acquiring data at wavelengths longer than gamma-ray wavelengths are usually configured to provide eventual output in image form. The most familiar electromagnetic sensor is the aerial camera. Its high resolution and simplicity is balanced by the limitations in spectral coverage. This ranges from the near ultraviolet through the visible range and on into the near-infrared.

The traditional approach to geological mapping has involved an on-the-ground or "boots on" search for rock outcrops. In most terrains, these are frequently scattered and isolated and are often fairly inaccessible. Mapping of large regions commonly requires years of fieldwork. The process can be

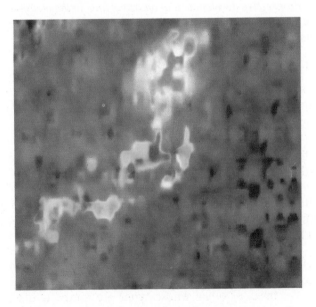

Figure 10-8 Computer map of rock exposures determined from gamma-ray spectroscopy (Western Geophysical Company of America)

accelerated and more economical if the geologist is provided with a series of aerial photographs that pinpoint outcrops and reveal structural associations. For many decades such photographs have served as the visual base from which maps are made by tracing the recognizable units where exposed. The geologist is then able to spot check the identity of each unit at selected localities and extrapolate the positions of the units throughout the photographs instead of surveying these units at many sites. While high-resolution aerial photographs are a prerequisite for detailed mapping, they have certain inherent limitations, such as vignetting, where the background tone varies outward from the centre, and geometric distortion. These distortions make the joining of overlapping views into mosaics to provide a regional picture of a large area difficult. Such synoptic overviews are particularly valuable in geology because the scales of interrelated landforms, structural deformation patterns and drainage networks are commonly expressed in tens to hundreds of kilometres, which is the range typically covered by satellite imagery. The chief value of satellite imagery to geological applications lies therefore in the regional aspect presented by individual frames and the mosaics constructed for vast areas extending over entire geological provinces. This allows, for example, whole sections of a continent subjected to glaciation to be examined as a unified surface on which different glacial landforms are spatially and genetically interrelated. Deposits of sands in desert environments can be tied to source areas, modifying surface obstructions and climatic influences.

LANDSAT imagery in particular has been found to be very useful for singling out linear features of structural significance. The extent and continuity of the faults and fractures are frequently misjudged when examined in the field or in individual aerial photographs. Even when displayed in mosaics of aerial photographs, these linear features are often obscured by differences in illumination and surface conditions that cause irregularities in the aerial mosaic. With satellite imagery, the trends of these linear features can usually be followed across diverse terrain and vegetation even though the segments may not be linked. In many instances these features have been identified with surface traces of faults and fracture zones that control patterns of topography, drainage and vegetation which serve as clues to their recognition. The importance of finding these features is that lineaments often represent major fracture systems responsible for earthquakes and for transporting and localizing mineral solutions as ore bodies at some stage in the past.

A major objective in geological mapping is the identification of rock types and alteration products. In general, most layered rocks cannot be directly identified in satellite imagery because of limitations in spatial resolution and the inherent lack of unique or characteristic differences in colour and brightness of rocks whose types are normally distinguished by mineral and chemical content and grain sizes. Nor is it possible to determine the stratigraphic age of recognizable surface units directly from remotely sensed data unless the units can be correlated with those of known age in the scene or elsewhere. In exceptional circumstances, certain rocks exposed in broad outcrops can be recognized by their spectral properties and by their distinctive topographic expressions. However, the presence of covering soil and vegetation tends to mask the properties favourable to recognition.

Figure 10-9 is a U-2 Thematic Mapper Simulator TM band 6-5-1 composite of the Darwin Hills, California. Data were acquired on 17 June 1983 by a NASA U-2 Earth Resources Observation Aircraft flying at 65,000 ft. Ground resolution is approximately 26 m. TM band 6 (thermal IR) is in red, TM band 5 (1.65 μm) is in green, and TM band 1 (0.5 μm) is in blue.

Bedded deposits of sulphide minerals in the Darwin Hills zone of hydro-thermal alteration have yielded in excess of 20 million ounces of silver, and over 65,000 tons of lead and antimony.

Figure 10-10 is a U-2 Thematic Mapper Simulator (TMS) image of the San Manuel porphyry copper deposit and the St. Anthony vein deposit, located 30 miles northeast of Tucson, Arizona. The latter produced nearly two million tons of silver, gold, vanadinite, molybdenite, lead, and zinc ores until its closure in 1955. The TMS 5-4-1 composite clearly shows much of the complex, highly faulted, geological structure of the area. The dark areas at the centre consist of indurated alluvial fans and latite flows. The lighter areas to the north are heavily fractured quartz monzonites and granodiorites, with diabase dikes appearing as dark striations. South-east of the darker area is the reddish-brown outcrop of the principal ore deposit, which consists of quartz monzonite intruded by granodiorite porphyry. The lower quarter of the image consists of partially indurated Plio-Pleistocene alluvial fans.

10.3.2 Geological information from the thermal spectrum

10.3.2.1 Thermal mapping

The application of thermal imagery in geological mapping is a direct result of the fact that non-porous rocks are better heat conductors than unconsolidated soils. At night, therefore, such rocks will conduct relatively more of the Earth's heat than the surrounding soil-covered areas, producing very marked heat anomalies which the scanner detects. Porous rocks, on the other hand, do not show the same intense heat anomalies on night imagery and after recent rainfall may actually produce cool anomalies due to their moisture content. The very strong heat anomalies produced by most rock types permit the detection of very small outcrops, the tracing of thin outcropping rock units and, depending on the nature of the soil, their detection below a thin soil cover. Loose sandy soils permit detection of suboutcrop below at least 20 cm of soil thickness but in moist clay soils there is virtually no depth penetration.

Figure 10-9 U-2 Thematic Mapper Simulator Image of Darwin Hills, California, with shaded areas showing alteration zones (NASA Ames Research Center)

Red .... Olivine basalt flows
Yellow/Yellow-Green .... Pyroclastic rocks, basalt cinders
Light Blue .... Quaternary alluvium
Dark Blue/Violet .... Limestone and dolomite
Green (Darwin Hills) .... Quartz monzonite
Turquoise (Darwin Hills) .... Alteration zone — silicated limestones, dolomites,
  shales (calc-silicates, calcarenite).

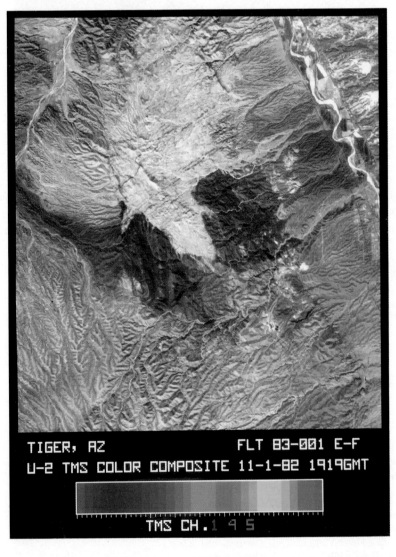

Figure 10-10 Thematic Mapper Simulator false-colour composite of bands 5, 4 and 1 (NASA Ames Research Center)

The identification of individual rock types is based mainly on field checking and on the basis of structure and texture, the latter being the result of the characteristic jointing that particular rock types exhibit. In theory, the effect of thermal inertia i.e. the rate at which particular rock types heat up or cool down during the night, or the employment of narrow-band infrared detectors can be used for the direct identification of rock type, but the practical application of these techniques is still under development.

10.3.2.2 Engineering geology

Figure 10-11 is a classic example of far-infrared imaging. The visible spectrum image on the left provides no indication of the buried stream channel appearing in the far-infrared image on the right. Not only is the buried stream course evident, but given the knowledge that the infrared image was acquired at night, certain inferences may be drawn. One is that the horizontal portion of the stream channel is probably of coarser sand and gravel than the vertical portion. This is because its bright signal suggests more readily flowing water, which is warmer than the night-air-cooled surrounding soil. The very dark signal adjoining and above the horizontal portion of the stream course, and to both sides of the vertical portion, probably represents clay, indicating moisture that cooled down many years ago, and remained cold compared with the surrounding soil because of poor permeability and thermal conductivity. Such imagery could be invaluable for investigating ground water, and avenues for pollution movement, exploration for placers, sand, gravel and clay, and emplacing engineering structures.

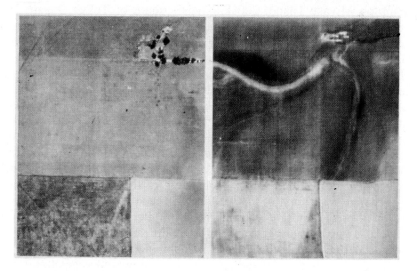

Figure 10-11 Visible (left) and thermal-infrared (right) image of a buried stream channel (Western Geophysical Company of America)

Unlike hydrogeological studies, or mineral exploration, engineering geological investigations are often confined to the upper 5 m of the Earth's surface. As thermal-infrared linescanning has proved to be very effective within the near-surface zone it is becoming increasingly useful in a variety of engineering geological problems such as those aimed at assessing the nature of material for foundations, excavations, construction materials and drainage purposes.

Density of materials and ground moisture content are the two dominant factors influencing tonal variations recorded in thermal imagery. Sub-surface solid rock geology may be interpreted from one or more of a number of "indicators", which include vegetation changes, topographic undulations, soil variations, moisture concentrations and mineralogical differences. As far as particle size or grading is concerned, in unconsolidated surface materials, lighter tones are caused by coarser gradings. Lighter tones also result from a greater degree of compaction of surface materials.

10.3.2.3 Geothermal and volcano studies

Geothermal mapping using scanner data is mainly carried out in geothermal energy exploration studies as part of investigations of alternative energy resources (Japanese Sunshine project) or for the monitoring of active volcanoes, particularly in the case of eruption prediction studies.

Figure 10-12 is an example of a ground-surface temperature map on which topographic contour lines are overlaid. The data are of Mount Oyama on Miyake Island, 200 kilometres south of Tokyo, Japan. Data were acquired and processed by the Asia Air Survey Company using a Daedalus AADS-1250 MSS and the Asia Digital Image Processing System (ADIPS) developed by the company.

Figures 10-13(a) and (b) represent a stereopair of the ground-surface temperature map generated by computer processing to produce an image of the temperature distribution on the stereo terrain model. Geothermal mapping sites are usually located in mountainous areas, and repetitive mapping of the same area is often required for monitoring purposes. Thus strict geometric rectification of the imagery is required. In the rectification process Digital Terrain Model (DTM) data provide information for relief displacement correction. Block adjustment of the MSS strips allows not only accurate estimation of the scanner's attitude and position, but also the creation of precise mosaics of multiple strips after radiometric correction of the data. Furthermore, terrain information such as slope, aspect and elevation derived from a DTM can be combined with the MSS data, and various effects due to solar radiation, soil moisture and vegetation may be evaluated for the refinement of the data.

A surface temperature map obtained from night-time thermal IR data is

Figure 10-12 A ground surface temperature map of Mount Oyama in Japan with topographic contours overlaid

primary data, calibration of which is difficult and usually carried out by using ground measurement data and the inner references of the MSS only. The principal difficulty is of evaluating the atmospheric effects for the night-time data as no standard reference exists. However, the path lengths between sensor and ground objects can be calculated from exterior orientation parameters and DTM data, enabling the relationship between relative differences of path length and temperature discrepancies to be established. Although the ground truth data have greater weight, analysis of path length effects yield better understanding of the atmospheric effects that

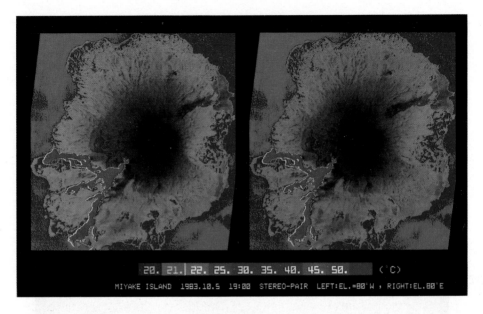

Figure 10-13 (a, left), (b, right) Stereopair of colour-coded temperature maps of Miyake Island (Asia Air Survey Company)

may vary locally in the survey area. Ground-surface temperature obtained from the thermal IR data is a result of a combination of heat flow from beneath the ground surface and contamination from other influences such as air temperature, solar heating etc. which may be filtered out if the data are used in conjunction with elevation, slope, and aspect derived from a DTM. Slope and aspect of the ground are the main components of the thermal contamination caused by topography and solar heating. The southward slope is usually warmer than the northward slope in the northern hemisphere and temperature differences are clearly seen even on predawn thermal IR data. The correction for topographic conditions may be as large as 1°C for night-time data.

Differences of emissivity between objects due to different land cover conditions are a further difficulty in the evaluation of thermal IR data. These differences have to be taken into account when the data are analyzed. A land cover map may be compiled on the basis of the multi-spectral characteristics of objects. Surface temperature may be obtained from non-vegetated areas identified in the land-cover map. Because of vegetation cover the ground-surface temperature data available is usually very sparsely distributed. If it is assumed that the overall distribution pattern of the ground-surface temperature reflects the underground temperature distribution, which is usually governed by the geological structure of the area, a trend-surface analysis may be applied to interpolate between the sparse ground temperature data and so make it possible to visualize the trend of the ground-surface temperature which is usually obscured by the highly frequent change in ground-surface temperatures. Of course, by combining the thermal IR data with other survey data even more useful information can be drawn and a better understanding of the survey area achieved.

The refined remote sensing data may then be cross-correlated by computer manipulation of multi-variate data sets including geological, geophysical and geochemical information. These remote sensing data, together with other remote and direct sensing measurements, may then be used to target drill holes to test the geothermal site for geothermal resources.

10.3.2.4 Detecting underground and surface coal fires

Self combustion is one of the many problems associated with coal mining. Incipient heating, both in the mines and in storage, at times gives rise to fires; these can be detrimental to the production and overall quality of coal and can constitute a major hazard, there being many instances of injuries and fatalities due to burns or poisoning by noxious gases. It is therefore important that zones of self combustion or fire be detected at the earliest possible time and, thereafter, continuously monitored until such time as suitable remedial action has been effected.

The traditional method for the monitoring of self combustion in coal is by the use of thermistors. Temperatures are measured at as many points as possible on the ground or dump, and an isothermal map constructed. From this information areas of relatively higher temperatures, which could relate to self combustion, can be delineated. This method is, however, somewhat conjectural particularly if the data points are sparse.

The infrared linescanner provides a very attractive alternative to the use of thermistors. It is capable of supplying a continuous record of very small temperature changes without having to come too close to the source itself. In the case of underground fires, heat produced from burning coal would not necessarily escape to the surface, but the ground above the fire will be heated by conduction and be indicated in the data by a thermal anomaly.

10.3.3 Geological information from radar data

Operating at wavelengths beyond the infrared region of the electromagnetic spectrum are the various radar devices. Radar supplies its own illumination and accordingly can collect data by day or by night. Because of the long wavelength used, radar can also generally collect data through cloud cover and is therefore invaluable for mapping in humid tropic environments, which are generally characterized by perpetual cloud cover.

Figure 10-14 is an X-band, synthetic aperture radar image of an arid

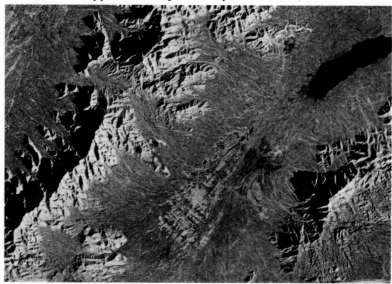

Figure 10-14 X-band SAR image of an arid environment in Arizona (Western Geophysical Company of America)

environment in Arizona. The radar, responding to surface cover texture and geometry, shows the brightest return signals from the bare rock mountain peaks and talus. The return signals decrease progressively downslope, through the bejada slopes, and on into the Bolson plain, with their possible playa features, which are totally specular. That is, they reflect the illuminating energy away from, rather than back to, the receiver. These "dark" signals are from the finest material in the area, probably clays and silts.

Stringers of bright signals may be observed in the Bolson plain. These are probably from coarse sands and gravels and represent the braided courses of the highest velocity streams entering the area.

Thus radar, in this sense, is useful in exploring for sand and gravel construction materials and placers, that is for emplacing construction on the good, coarse materials rather than on the clays and silts. Radar has been used for detecting and mapping many other environmental factors. A major product used for such work is the precision mosaic, from which geological maps, geomorphological maps, soil maps, ecological conservation maps, land use potential maps, agricultural maps and phytological maps have been generated.

Radar imaging geometry enhances lineaments. Lineaments are the terrain surface expression of fractures, jointing and other linear geological phenomena that occur anywhere from the terrain surface down to possibly great depths.

Alteration products associated with rocks and ore bodies are often widespread at the surface and may show diagnostic indications in satellite images. By ratioing the spectral information from LANDSAT MSS bands, information unnoticed in a single band or a black and white image, such as iron oxide surface stains overlying altered ore deposits, may be extracted.

10.3.4 Geological information from potential field data

The aeromagnetometer is the most generally used potential field sensor. Millions of line kilometres of aeromagnetic data have been acquired over land and sea, globally, since its inception. Figure 10-15 is included to indicate that digitally processed aeromagnetic data can provide information concerning the structural geology and lithology of the environments to a depth of many thousands of feet.

The "poor second sister" of the potential field sensors is the airborne gravity meter. A gravity meter can provide structural geological and lithological information from greater depths than can an aeromagnetometer, particularly in situations where the latter may be limited in its data acquisition, such as in the presence of power lines and metal fences, but its operational requirements are more demanding and costly.

10.3.5 Geological information from sonars

Sonar is the major sensor used in sea-floor mapping. Figure 10-16(a) is of sea-floor features at about 1200 metres depth in the Juan de Fuca area off the coast of Oregon, U.S.A., and depicts an environment of sea-floor spreading which is evident from the division of the volcanic cones that has taken place.

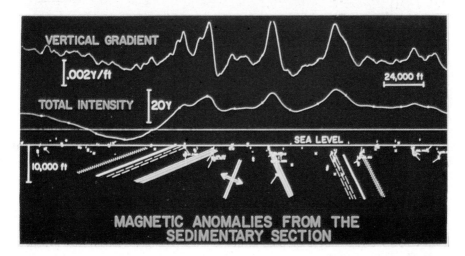

Figure 10-15 Structural geology factors revealed by aeromagnetic data (Western Geophysical Company of America)

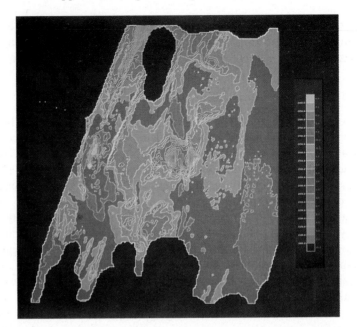

(a)

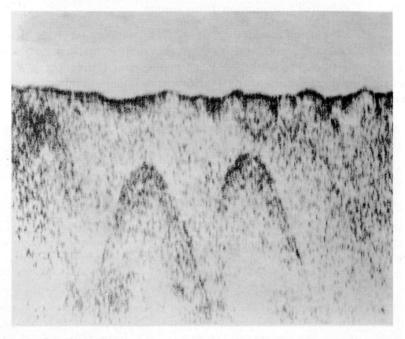

(b)

Figure 10-16 Sonar images showing (a) sea floor spreading at 2,000–2,500 m depths in the Gorda Rise, off the coast of Oregon (Western Geophysical Company of America) and (b) sub-bottom profiling of Callahan and Sumner tunnels under Boston Harbor, Mass. (Klein Associates Inc.)

Sub-bottom information is also obtainable from sonars. Figure 10-16(b) shows a profile of the Callahan and Sumner Tunnels under Boston Harbour, Massachusetts, U.S.A.

10.4 Applications to the biosphere

Even in the most technologically advanced countries, up-to-date and accurate assessments of total acreage of different crops in production, anticipated yields, stages of growth and condition (health and vigour) are often incomplete or non-timely in relation to the information needed by agricultural managers. These managers are continually faced with decisions on planting, fertilizing, watering, control of pests and disease, harvesting, storage, evaluation of crop quality and planning for new cultivation areas. Remotely sensed information is used to predict marketing factors, evaluate effects of crop failure, assess damage from natural disasters and to aid farmers in determining when to plough, water, spray or reap. The need for accurate and timely information is particularly acute in agricultural information systems because of the very rapid changes in the condition of agricultural crops and the influence of crop yield predictions on the world market; it is for these reasons that, as remote sensing technology has developed, the potential for this technology to be used has received widespread attention. Previously, aircraft surveys were used sporadically to assist crop and range managers in gathering useful data. With the advent of multi-spectral repetitive satellite imagery, the possibility of an automated crop inventory is now approaching reality.

Figures 10-17 (a,b,c) are colour-infrared (TMS-4,3,2), natural colour and water colour (TMS-3,2,1) images of San Francisco Bay and peninsula at ground resolution of 50 m obtained from a wide-angle Daedalus DS-1260 linescanner.

Colour-infrared film is sensitive to the green, red, and near-infrared (500 – 900 nm) portions of the spectrum and is widely used in aerial and space photographic surveys for land use and vegetation analysis. Living vegetation reflects light in the green portion of the visible spectrum to which the human eye is sensitive. Additionally, it reflects up to 10 times as much in the near-infrared (700 – 1100 nm) portion of the spectrum which is just beyond the range of human vision. When there is a decrease in photosynthesis, whether caused by normal maturation or stress, there is a corresponding decrease in near-infrared reflectance. Living or healthy vegetation appears as various hues of red in colour-infrared film. If diseased or stressed, the colour response will shift to browns or yellows due to the decrease in near-infrared reflectance. This film is also effective for haze penetration because blue light is eliminated by filtration.

Crops are best identified from computer-processed digital data that

(c)

(b)

(a)

Figure 10-17 Thematic Mapper Simulator (a) colour infrared, (b) natural colour and (c) water colour images of San Francisco Bay (NASA Ames Research Center)

represent quantitative measures of radiance. In general, all leafy vegetation has a similar reflectance spectrum regardless of plant or crop species. The differences between crops, by which they are separated and identified, depend on the degree of maturity and percentage of canopy cover, although differences in soil type and soil moisture may serve to confuse the differentiation. However, if certain crops are not separable at one particular time of the year, they may be separable more readily at a different stage of the season due to differences in planting, maturing and harvesting dates. The degree of maturity and the yield for a given crop also influence the reflectance at any stage of growth. This maturity and yield can be assessed as the history of any crop is traced in terms of its changing reflectances. When a crop is diseased or seriously damaged (for example, by hail) its reflectances decrease, particularly in the infrared, allowing the presence of stress to be recognized. Lack of available moisture also stresses a crop, the effect of which again shows up as a reduction of reflected light intensity in the infrared, usually with a concomitant drop in reflectance in the green and a rise in the red. Crop acreage estimation consists of two parts, the mensuration of field sizes and the categorization of those fields by crop type. The mensuration process can sometimes be facilitated by manipulating imagery to make field boundaries more distinct. The categorizations of those fields by crop type is then usually performed by multi-spectral classification (see Section 9.7). Similarly, the biomass, or amount of feed available in grasses, bush, and other forage vegetation of the rangeland, may also be estimated from measurements of relative radiance levels.

The three year Large Area Crop Inventory Experiment (LACIE), using LANDSAT MSS imagery, demonstrated that the global monitoring by satellite of food and fibre production was possible. The U.S. Department of Agriculture (USDA) is making the transition from being a partner with NASA and NOAA in the now concluded LACIE proof-of-concept programme to developing AgRISTARS (Agricultural Resource Inventories Through Aerospace Remote Sensing) as an operational test system geared to support new initiatives in crop forecasting. The AgRISTARS research and development is directed towards a second-generation crop assessment system with expanded capabilities. Where LACIE's mission was to prove the feasibility of LANDSAT for yield assessment of one crop, wheat, it is now extended to determine the feasibility of monitoring eight crops on a global scale.

With the U.S.A. becoming an increasing exporter of key foods, it has been estimated by LACIE participants that the economic benefits to the U.S. agricultural industry from improved estimates of foreign crop yields could be of the order of several million dollars annually, wheat alone accounting for a large part of this. The previous USDA system of developing foreign estimates from a variety of sources was not very reliable because many foreign producers lack accurate data sources. The LACIE experiment highlighted the potential impact that a credible crop yield assessment could have on world food marketing, administration policy, transportation and other related factors. In 1977, during Phase 3 of LACIE, it was decided to test the accuracy of Soviet wheat crop yield data by using LANDSAT-3 to assess the USSR's total production from early season to harvesting. In January 1977 the Soviet Union officially announced that it expected a total grain crop of 213.3 million metric tons. This was about 13% higher than the country's 1971 – 1976 average. Since Soviet wheat historically has accounted for 48% of its total grain production, the anticipated wheat yield would have been about 102 million metric tons for the year. LACIE computations, made after the Soviet harvests, but prior to the USSR release of figures, estimated Russian wheat production at 91.4 million metric tons. In late January 1978, the USSR announced that its 1977 wheat production had been 92 million metric tons. The USDA final estimate was 90 million metric tons. Previous USDA assessments of Soviet wheat yield had had an accuracy of 65/90, meaning that the USDA's conventionally collected data could have an accuracy of ±10% only 65% of the time. The LACIE programme was designed to provide a crop yield assessment accuracy of 90/90, or within ±10% in 90% of the years the system is used.

In forestry, the LANDSAT MSS have proved effective in recognizing and locating the broadest classes of forest land and timber and in separating deciduous, evergreen and mixed (deciduous-evergreen) communities. More discrete classifications are possible with data from the Thematic Mapper. Further possibilities include measurement of the total acreage given to forests and changes in these amounts. Timber volume, age of forests and presence of disease or pest infections can also be determined. Heat sensitive channels on the LANDSAT Thematic Mapper and the AVHRR also have application to forestry through their ability to detect fires. Figure 10-18 is a pseudo-colour thermal IR (U-2 TMS band 6) image of a forest fire in Cedar Grove, Kings Canyon National Park, California on 2 October, 1980 at 1200 PDT.

Also of interest to land cover studies is the ability to recognize the major soil associations and to discriminate the specific soil types. This is feasible wherever the soil possesses a distinctive spectral signature and/or supports characteristic, identifiable vegetation.

10.4.1 Spatial Information Systems, land use and land cover mapping

In recent years satellite data have been incorporated into sophisticated information systems of a geographical nature, allowing the synthesis of remotely sensed data with existing information.

The growing complexity of society has increased the demand for timely and accurate information on the spatial distribution of land and environmental resources, social and economic indicators, land ownership

Figure 10-18 Pseudo-colour thermal-infrared band Thematic Mapper Simulator image of a forest fire (NASA Ames Research Center)

and value, and their various interactions. Land and Geographic Information Systems (GIS) attempt to model, in time and space, these diverse relationships so that at any location, data on the physical, social and administrative environment can be accessed, interrogated and combined to give valuable information to planners, administrators, resource scientists and researchers. Land Information Systems are traditionally parcel based and concerned with information on land ownership, tenure, valuation and land use and tend to have an administrative bias. Their establishment is dependent on a thorough knowledge of the cadastral system and of the horizontal and vertical linkages that occur between and within government departments that collect, store and utilize land data. Geographic Information Systems have developed from the resource-related needs of society and are primarily concerned with inventory based on thematic information, particularly in the context of resource management (see Burrough, 1986; Rhind and Mounsey, 1991). Increasingly the term Spatial Information Systems is being used to describe systems that encompass and link both Land and Geographic Information Systems.

By their very nature Land and Geographic Information Systems rely on data from many sources such as field surveys, censuses, records from land title-deeds and remote sensing. The volume of data required has inevitably linked the development of these systems to the development of information technology with its growing capacity to store, manipulate and display large amounts of spatial data in both textural and graphical form. Fundamental to these systems is an accurate knowledge of the location and reliability of the data; accordingly remote sensing is able to provide a significant input to these systems, in terms of both the initial collection and subsequent updating of the data.

Land use refers to the current use of the land surface by man for his activities, while land cover refers to the state or cover of the land only. Current remotely sensed data of the Earth's surface generally provide information about the land cover, with interpretation or additional information being needed to ascertain the land use. Land use planning is concerned with achieving the optimum benefits for mankind in the development and management of the land for various purposes, such as food production, housing, urbanization, manufacture, supply of raw materials, power production, transportation and recreation. This planning aims to match land use with land capability and to tie specific uses with appropriate natural conditions so as to provide adequate food and materials supplies without significant damage to the environment. Land use planning has previously been severely hampered both by the lack of up-to-date maps, showing which categories are present and changing in an area or large region, and by the inadequacies of the means by which the huge quantities of data involved were handled. The costs involved in producing land cover, land use and land capability maps have prohibited their acquisition at useful working scales. Vast areas of Africa, Asia and South America remain poorly, and often incorrectly, mapped. A UNESCO-sponsored project has completed a series of land use maps at scales of 1:5,000,000 to 1:20,000,000. Although invaluable as a general record of land use (and for agriculture, hydrology and geology), these maps have insufficient detail to assist developers and managers in many of their decisions. Furthermore, frequent changes in land usage are difficult to plot at such small scales. Satellites are able to contribute significantly to the improvement of this situation.

Land use can be deduced or inferred indirectly from the identity and distribution patterns of vegetation, surface materials and cultural features as interpreted from imagery. With supplementary information, which may be extracted from appropriate databases or information systems but is usually simply the accumulated knowledge of the interpreter, specific categories of surface features can be depicted in map form. Accordingly, through both satellite and aircraft coverage as the situation requires, it is now possible to monitor changing land use patterns, survey environmentally critical areas and perform land capability inventories on a continuing basis. The repetitive coverage provided by current satellites allows the continual updating of information systems and maps, although the frequency of revision will be dependent on the scale of map involved and the geographical situation of the area in question.

10.5 Applications to the hydrosphere

10.5.1 Hydrology

The more information there is available about the hydrologic cycle, the better the water manager is able to make decisions with regard to the allocation of water resources for consumption, industrial use, irrigation, power generation, and recreation. In times of excess, flood control may become the primary task; in times of drought, irrigation and power generation may be the first concern. The perspective gained by satellite remote sensing adds the aerial dimension to the conventional hydrologic data collected at point measurement stations, see, for example, Figure 10-19 which is a simulated Thematic Mapper image of a section of the Igikpuk river on the north slope of Alaska, U.S.A. Estimates of the occurrence and distribution of water are greatly facilitated with medium-resolution satellite data, while the repetitive coverage provides a first step towards the assessment of the rapid changes associated with the hydrological cycle.

Fortunately, many of the hydrologic features of interest for improved water resource management are easily detected and measured with remote sensing systems. Although water sometimes reflects light in the visible wavelengths in a similar manner to other surface features, it strongly absorbs light in the near-infrared. As a consequence, standing water is very dark in the near-infrared, contrasting with soil and vegetation which both appear bright in this part of the spectrum. Thus, in the absence of cloud, surface water can easily be distinguished and monitored in the optical wavebands.

Snow depth and snow-covered area are two important parameters that determine the extent of water runoff in river basins after the snow has melted. In many parts of the world this runoff is important for drinking water supplies, hydro-electric power supply and irrigation. The NOAA polar-orbiting satellites cannot be used to determine snow depth, but they can be used to determine the snow-covered area in a river basin. Snow maps are produced by enlarging and rectifying a visible band image to match the selected river basin map. The river basin is then overlayed on the rectified satellite image. The analyst then traces the snow line from the

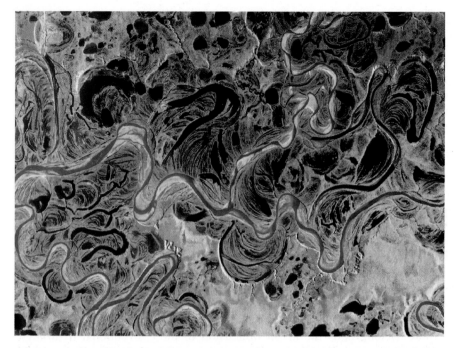

Figure 10-19 Simulated Thematic Mapper image of a section of the Igikpuk river on the north slope of Alaska (NASA Ames Research Center)

satellite image onto the appropriate basin map. Snow covered areas can be digitally enhanced, making it easy to compute or manually calculate the percent of snow-covered area in the basin. The river basin snow condition can then be monitored on a daily, weekly, or monthly basis.

In the visible region, snow obviously appears very bright, providing marked contrast with non-snow-covered surfaces. Discrimination between cloud and snow is not at all easy in the optical wavelengths. In regions where cloud cover presents a real problem, more sophisticated techniques incorporating mid-infrared, thermal-infrared or microwave wavelengths might be better suited to the task.

10.5.2 Oceanography and marine resources

Efficient management of marine resources and effective management of activities within the coastal zone is dependent, to a large extent, upon the ability to identify, measure and analyze a number of processes and parameters that operate or react together in the highly dynamic marine environment. In this regard, measurements are required of the physical, chemical, geometrical and optical features of coastal and open zones of the oceans. These measurements would include sea ice, temperature, current, suspended sediments, sea state, bathymetry and water and bottom colour. Different remote sensing capabilities exist for the provision of the required information involving one or a combination of measurement techniques.

10.5.2.1 Satellite views of upwelling

The phenomenon of wind-driven coastal upwelling, and the resulting high biological productivity off the west coast of North America, is dramatically revealed in these images of sea-surface temperature and chlorophyll pigments from NOAA and NASA satellites.

During the summer, northerly winds drive surface waters offshore and induce a vertical circulation which brings cooler water rich in plant nutrients to the sunlit surface. Microscopic marine algae, the phytoplankton, grow rapidly with the abundant nutrients and sunlight, initiating a rich biological food web of zooplankton, fish, mammals and birds. Such coastal upwelling regions support important fisheries around the world and are found off Peru and Ecuador, north and south West Africa, as well as off California and Oregon in the United States.

Figure 10-20 shows information derived from satellite observations made on 8 July 1981, during a period of sustained winds from the north. Figure 10-20 (a), derived from data obtained from the AVHRR on NOAA-6, shows the cool sea-surface temperatures along the coast (purple), especially noticeable at Cape Blanco, Cape Mendocino and Point Arena. The sea-surface temperature in the upwelling centres is about 8°C, as compared to 14°C further offshore. Several large eddies are visible, and long filaments of the cooler water meander hundreds of kilometres offshore from the upwelling bands in the California Current system.

Figure 10-20 (b) showing phytoplankton chlorophyll pigments was derived from data obtained from the CZCS on NIMBUS-7. The CZCS measures the colour of the ocean, which shifts from blue to green as the phytoplankton and their associated chlorophyll pigments become more abundant. Ocean colour measurements can be converted to pigment concentrations with a surprising degree of accuracy, and hence provide a good estimate of biological productivity using data obtained from space. This CZCS image shows the enhanced production along the coast due to the upwelling of the cool, high nutrient water. The image also shows the entrainment of the phytoplankton in the filaments of water being carried offshore, indicating that coastal production is an important source of biological material for offshore waters, which do not have a ready source of plant nutrients.

Data for these two scenes were taken over eight hours apart, and exhibit noticeable differences in cloud patterns (black and white regions) as a result of the time difference. Changes in sea-surface temperature and chlorophyll patterns also occurred, but are not so obvious because the ocean moves much more slowly than the atmosphere.

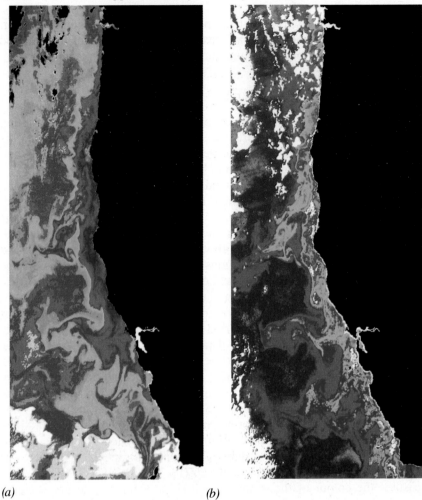

(a)   *(b)*

Figure 10-20 (a) Sea surface temperature determined using data from the AVHRR on the NOAA-6 satellite; and (b) the corresponding image of phytoplankton chlorophyll pigments made using data from the Coastal Zone Colour Scanner (CZCS) on the NIMBUS-7 satellite (NASA Goddard Space Flight Center). These computer images were produced by M. Abbot and P. Zion at the Jet Propulsion Laboratory. They used satellite data received at the Scripps Institute of Oceanography, and computer-processing routines developed at the University of Miami

Enhanced levels of chlorophyll pigments can be seen in regions where upwelling and temperature signals are not apparent, such as in the southward spreading plume of the Columbia River (northwest of Portland) and in the outflow of San Francisco Bay. These high levels are the result of the addition of nutrients from the rivers and estuaries. However, due to the suspended sediment content in these areas, the satellite data may be less accurate than elsewhere.

Until recently, the California Current was thought to be a broad, slow current flowing uniformly to the south. Analysis of satellite data has revealed a very complex system of swirls, jets and eddies, having only an average southerly flow. These two images serve to demonstrate the complexity of oceanic processes and, especially, show one aspect of the coupling between atmospheric processes (wind speed and direction), ocean circulations (upwelling and offshore transport) and the chemical and biological processes involved in marine ecosystems. They also illustrate the importance of satellite observation systems for increasing our understanding of large-scale oceanic processes.

10.5.2.2 Sea-surface temperatures

Figure 10-21 was acquired by the NOAA-7 AVHRR on 24 April 1982. The image includes approximately four minutes of data from 18:51:01 to 18:55:01 GMT and represents the heat emitted by the ocean surface in the thermal-infrared (channel 4, 10.5 µm to 11.5 µm). Ground resolution is approximately 1.1 km.

In general, light grey tones depict cool areas and dark tones depict warmer areas (except for the clouds which have been slightly enhanced for detail). The cooler slope and shelf waters along the east coast of the United States are lighter in tone, whereas the main core of the warmer Gulf Stream appears darker. Meanders and eddies (both warm and cold) are easily recognizable. Imagery such as this is used daily by NOAA oceanographers

Figure 10-21 AVHRR thermal-infrared image from NOAA-7 circa 1853 GMT on 24 April 1982 (NOAA)

to plot the course of the Gulf Stream and its Eddies. Data in the form of analyzed charts are provided daily to the fisheries industry and shipping concerns, while information on the location of the north wall of the Gulf Stream and the centre of each eddy is broadcast daily over the Marine Radio Network.

Since certain species of commercial and game fish are indigenous to waters of specific temperature, fisherman can save a great deal of money in fuel costs by being able to locate areas of higher potential. Because of the relatively strong currents associated with the main core and eddies, commercial shipping firms and even sailors can take advantage of these currents, or avoid them, and realize similar savings in fuel and transit time.

A temperature map obtained from SMMR data averaged over three days to provide the sea ice and ocean-surface temperature, spectral gradient ratio and brightness temperature over the polar region at 150 km resolution has already been given in Figure 2-15. Information on the sea-ice concentration, spectral gradient, sea-surface wind speed, liquid water over oceans, percent polarization over terrain and sea-ice multi-year fractions may also be obtained from the SMMR.

10.5.2.3 *Applications of lidars to bathymetry*

Trials of lidar bathymetry have been undertaken to determine the depths of the coastal waters and sheltered lagoons in the Ile de l'Est region of the Magdalen Islands in the Gulf of St. Lawrence. This is an excellent site for testing a lidar bathymeter as it is characterized by a wide range of water turbidities, bottom characteristics and sea states. Figure 10-22 shows a topographic map of the area of the Magdalen Islands. The northern islands of the group surround a shallow tidal lagoon, the bottom of which is covered with vegetation; water depths reach only about 8 m in the lagoon, but exceed 20 m in the water off the north coast.

The lidar beam attenuation coefficient was measured to be 1.7 m^{-1}. On

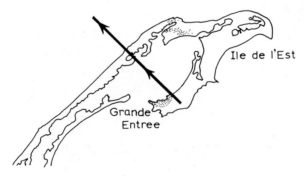

Figure 10-22 Lidar bathymeter flight line, 2 August 1979 Magdalen Islands (O'Neil et al., 1980)

the Gulf side of the islands, where the bottom is clean sand, the observed effective lidar attenuation coefficient was found to be in reasonable agreement with the measured beam attenuation coefficient; the point to point variation of the attenuation coefficient was quite small, with no appreciable variation with depth.

As the lidar bathymeter operates in a background limited mode, the upwelling sea-surface radiance was measured by gating on the green receiver photomultiplier tube (PMT) while the receiver was viewing the solar radiation reflected from the water. The resultant signal was then converted to a radiance value using the known sensitivity, collecting area and field of view of the receiver. The radiance value so calculated was $7 \times 10^{-5} W/(m^2 sr)$ which is within an order of magnitude of typical values found in the literature.

A total of 305 line-km were flown under favourable conditions of Sun angle and of weather. One of a total of thirty flight lines flown with the bathymeter is shown superposed on a topographic map of the Magdalen Islands in Figure 10-22. The laser was operated at 7 MW pulse power and its repetition rate was set to 9.3 Hz; the aircraft ground speed was approximately 75.7 ms⁻¹ (149 knots) for all flight lines. This resulted in an along-track depth sampling interval of approximately 8.1 m.

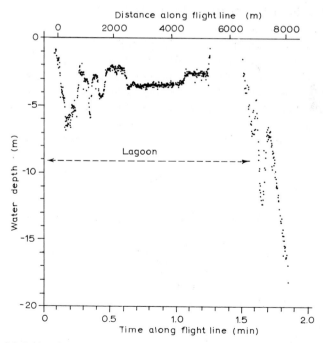

Figure 10-23 Lidar bathymeter trace from the flight line shown in Figure 10-22 (O'Neil et al., 1980)

Approximately 40,000 individual lidar returns were recorded during the mission.

Figure 10-23 is a plot of raw depths calculated from the lidar returns observed along the flight line shown in Figure 10-22. The horizontal scale of this plot is the distance along the line measured from an arbitrary starting point. The flight line began on the south shore of the islands, proceeded northwest across the lagoon and then into the north-coast waters. The plot illustrates the capability of the lidar bathymeter to achieve high resolution mapping of bottom contours. Several pronounced sand dunes that are visible in the lidar data were confirmed by the photography from this flight line. The water penetration capability of the sensor is also quite evident in Figure 10-23, where depths approaching 18 m are recorded; depths of slightly more than 20 m were successfully sounded on other flight lines. The minimum depth recorded on this flight line was 0.8 m. The uncertainty in the raw depth values is approximately 0.2 m.

These results were obtained without performing sophisticated processing of the lidar returns or making any assumptions concerning the scattering properties of the water. It is possible to correct the raw depths for the laser beam incidence angle and the aircraft attitude and position and it is further possible to apply tidal and sea-state corrections.

10.6 Applications to the cryosphere

Floating ice appears on many of the world's navigation routes for part of the year, while in the case of the high Arctic regions it is present for all of the year. It interferes with, or prevents, a wide variety of marine activities, including ships in transit, offshore resource exploration and transportation and offshore commercial fishing. In addition, it can be a major cause of damage, resulting in loss of vessels and equipment, loss of life and major ecological disasters. Accordingly, ice services are becoming increasingly available to marine users for a wide variety of applications. These include the navigation of vessels through the ice fields, the planning of ship movements and routings and of inshore and offshore fishing activities, the extension of operational shipping and offshore drilling seasons through forecasts of ice growth and break-up and the assistance of offshore drilling feasibility, economy and safety. These ice services have resulted in the reduction of maritime insurance rates and have contributed to the design of marine vessels and structures which are economical, yet safe. Current ice information charts providing daily up-to-date information on the position of ice edges, concentration boundaries, ice types, floe sizes and topographic features are prepared from ice data obtained from aircraft, satellites, ships and shore stations. Remote sensing techniques are particularly useful for gathering this ice information, both from aircraft and satellite platforms. The aircraft are specially equipped with transparent domes for visual observation, Side Looking Airborne Radar (SLAR) for all-weather information gathering capability and laser profilometers for accurate measurement of surface roughness.

Figure 10-24 is a SLAR image of an exploration platform in sheet ice surrounded by very clearly defined tracks left by an attendant icebreaker. The ice breaker is permanently on station in support of the exploration platform to break up the moving ice sheet before it interferes with the platform itself. It is therefore fairly easy to deduce the prevalent directions of ice flow. The resupply lanes are similarly obvious.

Figure 10-25 shows the mean monthly surface emissivity for January 1979 measured at 50.3 GHz as derived from the analysis of HIRS-2/MSU data. Sea ice extent and snow cover can be determined from this field. The emissivity of snow-free land is typically 0.9 to 1.0 while the emissivity of a water surface ranges from 0.5 to 0.65, increasing with decreasing surface temperature. Mixed ocean-land areas have intermediate values. The continents are clearly indicated as well as a number of islands, seas and

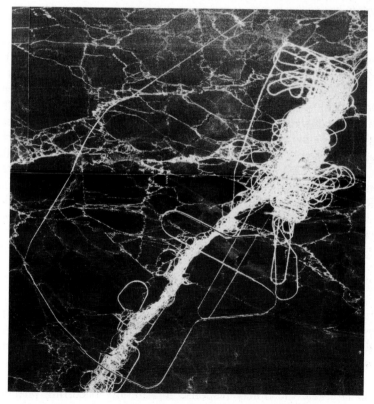

Figure 10-24 SLAR image of an icebreaker and drilling ship (Canada Center for Remote Sensing)

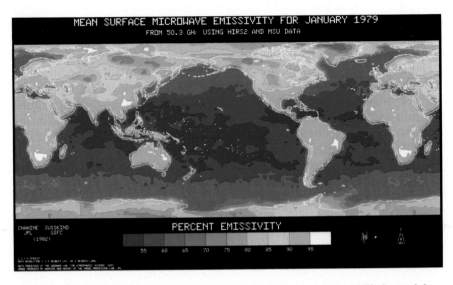

Figure 10-25 Mean monthly microwave emissivity for January 1979 derived from HIRS2/MSU data (supplied by NASA)

lakes. Snow covered land has an emissivity of 0.85 or less, with emissivity decreasing with increasing snow depth. The snow line, clearly visible in North America and Asia, gives good agreement with that determined from visible imagery. Newly frozen sea ice has an emissivity of 0.9 or more. Note for example Hudson's Bay, the Sea of Okhotsk, the centre of Baffin Bay, and the Chuckchi, Laptev and East Siberian Seas. Mixed sea ice and open water has emissivities between 0.69 and 0.90. The onset of significant amounts of sea ice is indicated by the 0.70 contour. Comparisons of this in Baffin Bay, the Denmark Strait and the Greenland Sea show excellent agreement with the 40% sea ice extent determined from the analysis of SMMR data from the same period. Multi-year ice, such as found in the Arctic Ocean north of the Beaufort Sea, is indicated by emissivities less than 0.80.

10.7 Environmental applications

Remote sensing provides information related in some way or other to the quality, protection and improvement of land and water resources. At the same time information is obtained about man's effect on the environment, allowing monitoring to take place where this may be required. Of course not all adverse effects on the environment are due to man. Remote sensing can also facilitate the timely response to naturally occurring phenomena such as volcanic eruptions, earthquakes, hurricanes, tornadoes and forest fires.

10.7.1 Thermal imagery

Examples of the use of thermal imagery for environmental monitoring include investigations of heat loss from buildings, septic tank seepage into water supply and sewage outfall.

Figure 10-26 is a thermal-infrared image of the city of Dundee, Scotland. Light grey tones depict warm areas and dark tones depict cool areas. Poorly insulated buildings, with high heat loss, are brighter than those buildings with efficient insulation. Clearly, thermal imagery has direct application to energy conservation programmes. An image from the same survey was shown previously in Figure 2-4. It can be seen that the sewage dispersion is not particularly effective in the prevailing conditions. Since this is a thermal image the tail-off with distance from the outfall is possibly more a measure of the rate of cooling than the dispersal of the sewage.

The detection of contrasting thermal signals was useful for a cattle survey made in west Texas one summer. In this case daylight visible spectrum imaging proved useless as cattle would congregate beneath any available tree canopy to escape direct sunlight, but at night the cattle would move out onto the grasslands to graze. In the far-infrared surveys flown at night, the warm bodies of the cattle readily contrasted with the cool night air and background pasture enabling a direct count of the cattle population.

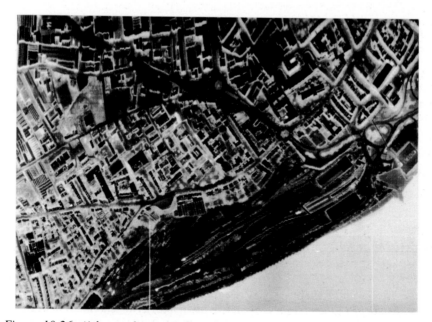

Figure 10-26 Airborne thermal-infrared survey data of part of the City of Dundee (Dundee University)

10.7.2 Air quality

The Stratospheric Aerosol Measurement (SAM II) instrument on the NIMBUS-7 satellite provides a measurement of the optical depth of the atmosphere and accordingly is able to provide information on air quality. Figure 10-27 charts several years of SAM II optical depth data at a wavelength of 1.0 μm for the atmosphere above the Arctic and Antarctic regions, which are nominally pollution free. However, major volcanic eruptions, as indicated by arrows, can be seen to disperse particulate matter even to these regions, the effect of which is to increase the optical depth.

Because of their global and frequent coverage, the NOAA polar-orbiting satellites are excellent platforms for observing volcanic eruptions, especially in remote parts of the world. The multi-spectral polar-orbiting data can be used to determine the vertical and horizontal plume morphology, track the ash cloud trajectory, monitor the eruption growth rate and calculate the plume altitude.

By comparing the satellite-derived temperatures to nearby radiosonde data (providing temperature versus height information), it is possible to determine the maximum altitude of the high-level plume. Figure 10-28 clearly shows the plume arising from the eruption of Mt. Etna, Sicily, in May 1983. The satellite-observed plume temperature was -33°C, corresponding to a height of 5 km.

In addition to their use in volcanological studies, satellite data may be

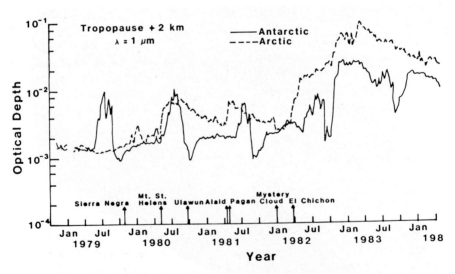

Figure 10-27 Stratospheric Aerosol Measurement (SAM II) optical depth data at a wavelength of 1·0 μm for the atmosphere above the Arctic and Antarctic regions. Major volcanic eruptions are indicated by arrows (NASA Goddard Space Flight Center)

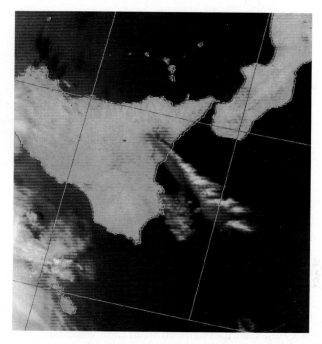

Figure 10-28 Smoke plume from the eruption of Mount Etna volcano, Sicily, NOAA-11 AVHRR band-2 image, 1340 GMT on 15 May 1983 (Dundee University)

useful in issuing warnings to aircraft flying near erupting volcanoes. In June and July of 1982, two Boeing 747 aircraft flew into ash clouds from the erupting Galunggung volcano in Java. Both suffered multiple engine failures and severe losses of altitude. Fortunately, safe emergency landings were made with no injuries. These near-fatal incidents have prompted the International Civil Aeronautics Organization to examine the possibility of using satellite data to provide ash cloud warnings to pilots when a volcano erupts.

10.7.3 Pollution by hydrocarbons

Oil pollution though pipeline leakage and accidental spillage is an all too familiar example of the deleterious effect man can have on the environment. For oil pollution monitoring from satellites the spatial resolution is often a serious problem, although under favourable conditions it is possible to observe and monitor oil slicks using this data. For example, in the case of the IXTOC-1 blowout in the Gulf of Mexico (June – August 1979) evidence of the oil slicks resulting from the spill can be seen in many images from scanners on several satellites, namely LANDSAT MSS, AVHRR, CZCS and the scanner on the geostationary satellites GOES-E, an example from the last being shown in Figure 10-29. The United States

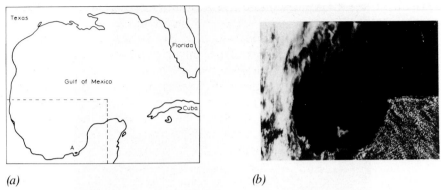

(a)   *(b)*

Figure 10-29 IXTOC-1 blowout: (a) sketch map of the Gulf of Mexico showing the position of the IXTOC-1 well at point A; and (b) band-1 (visible) AVHRR image, 2046 GMT on 6 June 1979

Coast Guard did indeed make some use of satellite remote sensing data in monitoring the movement of the oil originating from this spill. But this was an enormous spill which lasted for several months. Most other spills are much smaller and the oil involved disappears much more quickly. The United States Coast Guard concluded from their experience with IXTOC-1 that meteorological satellites were useful in the early stages, in the absence of all other sources of information, in providing a general sense of the magnitude of the spill and of the general direction of drift of the spilt oil. However, the low resolution and small scale and high distortion of the standard image product were felt to limit these remote sensing systems as reconnaissance tools. In practice the oil only seemed to be detectable with the visible sensors and not by the infrared sensors. During the time of the blowout both the LANDSAT-2 and LANDSAT-3 satellites were operating and a considerable amount of relevant imagery was obtained from them; it was found to be useful in corroborating the results from aircraft reconnaissance and in helping to establish the overall picture of the situation. But the infrequency of satellite overpasses, the problem of cloud cover, the rather low spatial resolution, and the lack of real-time processing and delivery of imagery prevented the use of LANDSAT for tactical response by the on-scene coordinator of the U.S. Coast Guard. In the event the use of low altitude visual aircraft reconnaissance backed up by an all-weather airborne remote sensing system (where weather and visibility required it) proved to be the most effective tactical and, along with accurate oil drift computer model forecasting, strategic aid for environmental pollution response planning for the IXTOC-1 spill.

Since a great deal of 'active' optical remote sensing work is related to the detection and monitoring of oil pollution, the ISOWAKE proposals put forward by the NATO Committee on the Challenges of Modern Society (CCMS) should not go unmentioned. These proposals involve the definition of a standard target to be used for the examination and evaluation of the performance of remote sensing systems being developed for oil pollution monitoring. The proposals also involve the specification of a recommended set of parameters of a controlled oil slick that should be measured to assist in the interpretation of remotely sensed data when systems are under development or evaluation and are being calibrated. These guidelines incorporate the use of conventional photography as well as other forms of remote sensing imagery but, while amplifying information was included on the use of photographic evidence, details were not presented on other forms of remote sensing.

Many different types of electronic sensors have detected illegal oil discharges. However, the degree and conditions under which these sensors can detect pollutant within acceptable levels of accuracy is uncertain. Therefore the interpretation of this imagery could be seriously questioned if it were to be presented as legal evidence. Within the framework of a NATO/CCMS Pilot Study on the Remote Sensing of Marine Pollution, nations participating agreed to conduct experiments to establish the appearance of wakes of ships containing known quantities of petroleum oils when viewed by various remote sensing devices. The purpose was to focus their efforts on determining the capability of existing electronic sensors to detect illegal discharges of oil. The ideal approach to accomplish this would be to initiate a series of international cooperative oil-spill experiments where each nation's oil detection sensors are flown simultaneously over the same discharge. These experiments would continue until a wide range of environmental conditions and oil discharge rates were recorded. Unfortunately, the logistical complexities of such an effort were beyond the means of the study.

As an alternative, the United Kingdom has proposed a single experiment that can be performed individually by each of the participating nations to test the performance of their remote sensing systems. A standardized oil wake discharge is defined to such a degree that it can be duplicated at any time and place convenient to the participating nation. The oil discharge experiment is designed to simulate real ballast water discharges. In addition, the extent to which each nation wishes to explore the oil detection phenomena is left to their individual needs and finances.

10.7.4 Applications of fluorosensing

The fluorescence spectra of hydrocarbons are quite distinct from those of most other materials found in the marine environment. An examination of a number of optically thick samples of various oils (Rayner, 1979) showed systematic differences in the fluorescence spectra, which generally exist as a single, broad, featureless peak covering most of the visible spectrum. Table 10-2 summarises the variation in the peak emission wavelength, λ_{max}, and the fluorescence conversion efficiency, n_{max}, at λ_{max},

Table 10.2   Fluorescence characteristics of oil classes

Oil type	λ_{max} (nm)	n_{max} (nm^{-1})
Light refined product	400	5×10^4
Crude oil	490	$1 \cdot 7 \times 10^4$
Heavy residual fuel	560	4×10^5

Excitation wavelength $\lambda_L = 337$ nm
(after Rayner, 1979)

for the three classes of oil when excited at a wavelength, λ_L, of 337 nm.

Table 10-2 can only be used to indicate general trends as there is a wide variation in physical and fluorescence properties of oils in all three classes. For example, there are heavy or weathered crude oils very similar to Bunker C, a residual product; whereas there are other fresh, light crudes that have fluorescence characteristics similar to marine diesel fuel (a light refined product).

In most oil spill situations, the oil has been observed to fluoresce much more strongly than the natural water background and there is little difficulty in detecting the oil. The identification of the oil may, however, be more difficult. In November 1978, a series of flight trials of the laser fluorosensor described in Section 5.3 were carried out over three spills off the coast of New Jersey. The spills were also observed with a large number of other airborne sensors thought to be of value for the detection, identification, classification, mapping and tracking of oil spills. Rhodamine WT dye, Merban crude oil (a light crude), and La Rosa crude oil (a heavy crude) were spilled on the sea surface in the course of the flight trials. The fluorescence emission spectra of these three target substances were compared in Figure 5-6. Typical fluorescence spectra from these three spills are shown in Figure 10-30.

Identification of spilt oil can be attempted through the comparison of the emission spectrum of the oil with the fluorescence signatures of known oils as one crosses the fluorescence anomaly.

10.7.5 Side scan sonar

Figure 10-31 shows one of the two remaining Wellington bombers in about 200 feet of water at the bottom of Loch Ness, Scotland. The twin-engined plane, Bomber N2980, R for Robert, a veteran of 14 raids over Germany, was ditched during a snow storm on New Year's Eve in 1940. Only one other of the 11,461 Wellingtons built remains, so a determined but unsuccessful attempt was made to raise the aircraft in September 1985. Group Captain Marwood-Elton, then a squadron leader, who crash landed the 14-ton aircraft, said that the landing was "perfectly easy and normal".

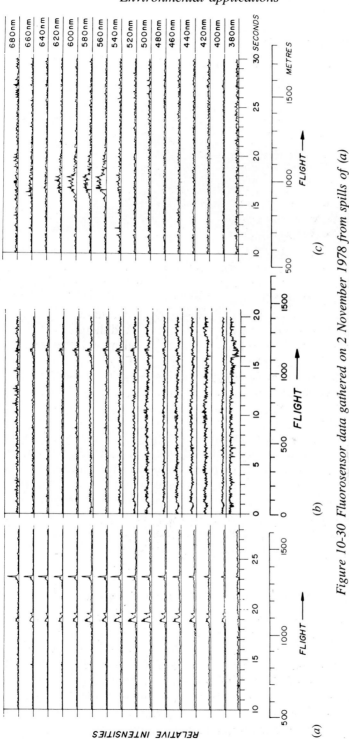

Figure 10-30 Fluorosensor data gathered on 2 November 1978 from spills of (a) Merban crude oil; (b) La Rosa crude oil; and (c) rhodamine WT dye (O'Neil et al., 1980)

Figure 10-31 Side-scan sonar image of a Wellington Bomber in Loch Ness, Scotland (Klein Associates, Inc.)

He and his co-pilot paddled ashore in a dinghy, the other six crew having baled out. Unfortunately one was killed when his parachute failed to open.

Side scan sonar provides a particularly effective means of locating wrecks at the bottom of the ocean. The successful location of the Titanic is one of the more well known examples. Since side scan sonar is an active technique protuberances on the seabed will cast a sonar shadow, which, depending on the angle of scan, may contain information not immediately evident from the direct echo from the object in view.

10.8 Applications of data collection systems

10.8.1 Meteorology

The 1978-1979 First Global Atmospheric Research Programme (GARP) Global Experiment (FGGE) successfully demonstrated the capacity of the ARGOS system to collect worldwide observations concerning the state of the atmosphere. The two main objectives of the FGGE were

1. To obtain a better understanding of atmospheric motion for

the development of more realistic models for extended range forecasting, general circulation studies and climate; and

2. To assess the ultimate limit of predictability of weather systems.

During the FGGE period, 301 buoys were deployed in the Southern Hemisphere south of 20 degrees South, 33 buoys in the equatorial region, 5 buoys in the North Atlantic and 24 buoys in the Arctic region. The buoy data were compatible with the ARGOS system (see Section 1.5.2). In addition to the buoys, the lifetime of which exceeded expectations, a tropical constant-level balloon system was implemented during the FGGE to obtain upper tropospheric data from the tropical region above the levels covered by the aircraft programme. One hundred and fifty-three balloons were launched during the first period and 157 during the second. In addition, the Balsamine experiment was performed during the period May – July 1979, as part of the Arabian Sea Phase of the Monsoon Experiment (MONEX) programme. This experiment, which concerned the monsoon circulation, used 88 constant-level superpressure balloons equipped with temperature, humidity, and pressure sensors. The ARGOS system located and collected the meteorological data from these balloons.

Meteorological users generally set up four types of measurement networks

1. Drifting buoys;

2. Ship terminals;

3. Automatic weather stations for obtaining data from isolated or inaccessible areas; and

4. Buoys moored in the open sea and transmitting real-time data.

The ARGOS System is used for all these situations.

10.8.2 Oceanography

Oceanography represents a variety of potential applications for the ARGOS system. Examples include

- Major international cooperation experiments such as the FGGE, already mentioned, and the European Communities' COST 43. The latter consisted in the setting up of a European experimental network of ocean-based stations in the North Atlantic. These stations provided real-time oceanographic and meteorological data.

- Major ocean current studies of the influence of such currents both on the general circulation of the oceans and the weather as well as on modelling and predicting marine pollution dispersal.

- Subsurface studies involving the collection of water temperatures at various depths, drogue-line parameters, acoustic parameters and current directions and speeds.

- The formation and movement of solitons (solitary waves) and of gyres produced by ocean currents.

- Ocean waves and swells (formation, spectrum, propagation, correlation with underwater ambient noise levels etc.).

Various aspects of the oceanographic applications of the ARGOS system are of use to the fishing industry, the results obtained being used to improve yields and to rationalize fish resource exploitation and management.

10.8.3 Investigations of the cryosphere

Glaciology is a field where site remoteness and severe environmental conditions have traditionally hampered data collection, especially in the Arctic and the Antarctic, where floating ice, i.e. icebergs, ice floes, etc., is studied and monitored. With the platform location capability of ARGOS, glaciologists are able to position and track platforms and hence, among other things, to study polar currents and compare buoy and iceberg courses. Present iceberg data collection stations measure atmospheric pressure, wind speed and direction, air and snow temperature, iceberg heading and tilting motions and iceberg surface strain.

Operational avalanche risk forecasting is a further application of the ARGOS System. The field stations developed for this type of application measure snow depth using an ultrasonic sensor, wind speed, air temperature and the vertical temperature profile. Satellite-based data collection is a particularly convenient way of acquiring data from such sites since avalanche-susceptible terrain generally prohibits the use of line-of-sight radio links.

10.8.4 Some other applications

The ARGOS System is finding a steadily increasing use in the offshore oil industry, in hydrological, seismological and volcanological studies and in the tracking of terrestrial and marine animals.

The ARGOS System has also been used to track competitors in a number of ocean yacht races. The utilization of the ARGOS System during these races, and especially the use made by the media of the data concerning competitors' positions, has heightened public interest and also considerably altered the spirit of such trans-ocean races. Rescues performed following mishaps during yacht races have demonstrated the usefulness of the ARGOS System for search and rescue. Accordingly, a special SARGOS (Search and Rescue ARGOS) System is being developed for this purpose using beacons that will transmit on a special frequency (406 MHz), with transmission only being activated in the event of an emergency.

10.9 Postscript

Since the launch of LANDSAT-1 in 1972 there has been a continuous and growing stream of satellite-derived Earth resources data available. It is a virtual certainty that tremendous amounts of additional remote sensing data will become available. The extent to which the data will actually be analyzed and interpreted for solving "real-world" problems is somewhat less certain. There is an acute shortage of suitable investigations which interpret and utilize the information to advantage because investment in the hardware systems involved in producing the data has not been matched with a similar investment in the use made of it. Remotely sensed data have been used extensively in research programmes although so far, space-acquired remote sensing data is being utilized much less in routine Earth resources investigations than was predicted in early optimistic estimates. Indeed, the short history of remote sensing has been one of transition from a total research orientation to the present emphasis on operational and quasi-operational programmes. Users have developed applications at their own pace, and the transition of these applications from a research to an operational orientation has been very gradual. However, impediments to the acceptance and development of remote sensing that once existed, such as the difficulties in handling the volumes of data remote sensors could generate and the limitation in the precision of measurements possible with data acquired by systems generally distant from their objectives, have now largely been overcome by advances in computing which have served to alleviate many of the data volume and manipulation problems.

The magnitude and complexity of the problems facing the world require coordinated planning, often in the regional context. Remote sensing has now made it possible for countries to obtain otherwise unavailable resource data to assist in the planning of their economic and social development, and may be seen accordingly to be of particular advantage to developing countries. For the present, however, it can be observed that most remote sensing effort is to be found in those parts of the world where computing and associated information technologies are already well developed. Perhaps more important than the ways different countries find to use remotely sensed information is the glamour attached to its use. All countries therefore seem poised to expand substantially their use of remote sensing data which is improving in both quality and diversity. It is to be hoped that this information can lead to the improvement of the quality of life of all who live on Earth.

It is important, as far as it is possible, to develop techniques that are capable of handling remotely sensed data in real time or very-near real time. When data are simply dumped in an archive with the intention that it should be used at a later date, experience seems to suggest that only a small fraction of the data is ever actually used.

References

Ahern, F. J. and Murphy, J., 1978, Radiometric calibration and correction of Landsat 1, 2 and 3 MSS data. Research Report 78–4, November 1978. Canada Centre for Remote Sensing (Energy, Mines and Resources), Ottawa, Canada

Alfoldi, T. T. and Munday, J. C., 1978, Water quality analysis by digital chromaticity mapping of Landsat data. *Canadian Journal of Remote Sensing*, **4**, 108

Anding, D. and Kauth, R., 1970, Estimation of sea surface temperature from space. *Remote Sensing of the Environment*, **1**, 217

Aranuvachapun, S. and Leblond, P. H., 1981, Turbidity of coastal water determined from Landsat. *Remote Sensing of the Environment*, **11**, 113

Barrett, E. C. and Curtis, L. F., 1982, *Introduction to Environmental Remote Sensing* (London: Chapman and Hall)

Barrick, D. E., 1971, Theory of HF and VHF propagation across the rough sea. 1, The effective surface impedance for a slightly rough highly conducting medium at grazing incidence. *Radio Science*, **6**, 517; Theory of HF and VHF propagation across the rough sea. 2, Application to HF and VHF propagation above the sea, *ibid*, 527

Barrick, D. E., 1972a, First-order theory and analysis of MF/HF/VHF scatter from the sea. *IEEE Transactions on Antennas and Propagation*, **AP-20**, 2

Barrick, D. E., 1972b, Remote sensing of sea state by radar. In *Remote Sensing of the Troposphere*, edited by V. E. Derr (Washington, DC: US Government Printing Office)

Barrick, D. E., 1977a, The ocean waveheight nondirectional spectrum from inversion of the HF sea-echo Doppler spectrum. *Remote Sensing of the Environment*, **6**, 201

Barrick, D. E., 1977b, Extraction of wave parameters from measured HF radar sea-echo Doppler spectra. *Radio Science*, **12**, 415

Barrick, D. E., Evans, M. W. and Weber, B. L., 1977, Ocean surface currents mapped by radar. *Science*, **197**, 138

Barrick, D. E. and Weber, B. L., 1977, On the nonlinear theory for gravity waves on the ocean's surface. Part II. Interpretation and applications. *Journal of Physical Oceanography*, **7**, 11

Baylis, P. E., 1981, Guide to the design and specification of a primary user receiving station for meteorological and oceanographic satellite data. In *Remote Sensing in Meteorology, Oceanography and Hydrology,* edited by A. P. Cracknell (Chichester: Ellis Horwood) p. 81

Baylis, P. E., 1983, University of Dundee Satellite Data Reception and Archiving Facility. In *Remote Sensing Applications in Marine Science and Technology,* edited by A. P. Cracknell (Dordrecht: D. Reidel) p. 29

Bernstein, R. L., 1982, Sea surface temperature estimation using the NOAA-6 satellite Advanced Very High Resolution Radiometer. *Journal of Geophysical Research,* **87C**, 9455

Bowers, D. G., Crook, P. J. E. and Simpson, J. H., 1982, An evaluation of sea surface temperature estimates from the AVHRR. *Remote Sensing and the Atmosphere: Proceedings of the Annual Technical Conference of the Remote Sensing Society,* Liverpool, December 1982 (Reading: Remote Sensing Society) p. 143

Bristow, M. and Nielsen, D., 1981, Remote monitoring of organic carbon in surface waters. Report No. EPA-600/4–81–001, Environmental Monitoring Systems Laboratory, US Environmental Protection Agency, Las Vegas, Nevada

Bristow, M., Nielsen, D., Bundy, D. and Furtek, R., 1981, Use of water Raman emission to correct airborne laser fluorosensor data for effects of water optical attenuation. *Applied Optics,* **20**, 2889

Bullard, R. K., 1983a, Land into sea does not go. In *Remote Sensing Applications in Marine Science and Technology,* edited by A. P. Cracknell (Dordrecht: D. Reidel) p. 359

Bullard, R. K., 1983b, Detection of marine contours from LANDSAT film and tape. In *Remote Sensing Applications in Marine Science and Technology,* edited by A. P. Cracknell (Dordrecht: D. Reidel) p. 373

Bullard, R. K. and Dixon-Gough, R. W., 1985, *Britain from space: an atlas of Landsat images* (London: Taylor & Francis)

Burrough, P. A., 1986, *Principles of Geographical Information Systems for Land Resources Assessment* (Oxford: Oxford University Press)

Callison, R. D. and Cracknell, A. P., 1984, Atmospheric correction to AVHRR brightness temperatures for waters around Great Britain. *International Journal of Remote Sensing,* **5**, 185

Cassanet, J., 1981, La télédétection HCMM et son application au littoral. *Memoires, Laboratoire de Géomorphologie, Ecole Pratique des Hautes Etudes,* **34**

Chedin, A., Scott, N. A. and Berroir, A., 1982, A single-channel double-viewing angle method for sea surface temperature determination from coincident METEOSAT and TIROS-N radiometric measurements. *Journal of Applied Meteorology,* **21**, 613

Colwell, R. N. (Editor), 1983, *Manual of Remote Sensing.* 2 Volumes. (Falls Church, VA: American Society of Photogrammetry and Remote Sensing)

Cook, A. F., 1985, Investigating abandoned limestone mines in the West Midlands of England with scanning sonar. *International Journal of Remote Sensing,* **6**, 611

Cott, P. J., 1980, The uses of satellites. III. Satellite communications. *Royal Society of Arts Journal,* **128**, 828

Cracknell, A. P., 1980, *Ultrasonics* (London: Taylor & Francis)

Cracknell, A. P., MacFarlane, N., McMillan, K., Charlton, J. A., McManus, J. and Ulbricht, K. A., 1982a, Remote sensing in Scotland using data received from satellites. A study of the Tay Estuary region using Landsat multispectral scanning imagery. *International Journal of Remote Sensing,* **3**, 113

Cracknell, A. P. and Singh, S. M., 1980, The determination of chlorophyll-a and suspended sediment concentrations for EURASEP test site, during North Sea Ocean Colour Scanner experiment, from an analysis of a Landsat scene of 27th June 1977. *Proceedings of the 14th Congress of the International Society of Photogrammetry,* Hamburg 1980, *International Archives of Photogrammetry,* **23**(B7), 225

Cracknell, A. P. and Singh, S. M., 1981, Proposed use of System ARGOS with data buoys for calibration of thermal infrared imagery of North British waters for sea surface temperature maps. *Proceedings of ARGOS System Workshop,* Bergen, 3–4 March 1981 (Toulouse: System ARGOS)

Cracknell, A. P., Storey, B. E. and Cracknell, C. P., 1982b, Low-cost digital image processing and display systems for teaching purposes. In *Remote Sensing and the Atmosphere: Proceedings of the Annual Technical Conference of the Remote Sensing Society,* Liverpool, December 1982 (Reading: Remote Sensing Society) p. 22

Crombie, D. D., 1955, Doppler spectrum of sea echo at 13.56 Mc/s. *Nature,* **175**, 681

Cutrona, L. J., Leith, E. N., Porcello, L. J. and Vivian, W. E., 1966, On the application of coherent optical processing techniques. *Proceedings of the IEEE,* **54**, 1026

Gordon, H. R., 1978, Removal of atmospheric effects from satellite imagery of the oceans. *Applied Optics,* **17**, 1631

Guymer, T. H., 1987, Remote sensing of sea-surface winds. In *Remote Sensing Applications in Meteorology and Climatology,* edited by R. A. Vaughan (Dordrecht: D. Reidel) p. 327

Guymer, T. H., Businger, J. A., Jones, W. L. and Stewart, R. H., 1981, Anomalous wind estimates from the SEASAT scatterometer. *Nature,* **294**, 735

Hayes, L. W. B., 1985, The current use of TIROS-N series of meteorological satellites for land-cover studies. *International Journal of Remote Sensing,* **6**, 35

Hoge, F. E. and Kincaid, J. S., 1980, Laser measurement of extinction coefficients of highly absorbing liquids. *Applied Optics,* **19**, 1143

Hoge, F. E., Swift, R. N. and Frederick, E. B., 1980, Water depth measurement using an airborne pulsed neon laser system. *Applied Optics,* **19**, 871

Hoge, F. E. and Swift, R. N., 1980, Oil film thickness measurement using airborne laser-induced water Raman backscatter. *Applied Optics,* **19**, 3269

Hoge, F. E. and Swift, R. N., 1981, Absolute tracer dye concentration using airborne laser-induced water Raman backscatter. *Applied Optics,* **20**, 1191

Holyer, R. J., 1984, A two-satellite method for measurement of sea surface temperature. *International Journal of Remote Sensing,* **5**, 115

Horvarth, R., Morgan, W. L. and Stewart, S. R., 1971, Optical Remote Sensing of Oil Slicks: Signature Analysis and Systems Evaluation. Final Report, US Coast Guard Project 724104.2/1

Hotelling, H., 1933, Analysis of a complex of statistical variables into principal components. *Journal of Educational Psychology,* **24**, 417, 498

Jarrett, O., Jr., Brown, C. A., Jr., Campbell, J. W., Houghton, W. M. and Poole, L. R., 1979, Measurement of chlorophyll-a fluorescence with an airborne fluorosensor, *Proceedings of the 13th International Symposium for Remote Sensing of Environment,* Ann Arbor, MI, April 1979, p. 703

Jones, W. L. and Schroeder, L. C., 1978, Radar backscatter from the ocean: dependence on surface friction velocity. *Boundary Layer Meteorology,* **13**, 133

Jones, W. L., Black, P. G., Boggs, D. H., Bracalente, E. M., Brown, R. A., Dome, G., Ernst, J. A., Halberstam, I. M., Overland, J. E., Peteherych, S., Pierson, W. J., Wentz, F. J., Woiceshyn, P. M. and Wurtele, M. G., 1979, SEASAT scatterometer: results of the Gulf of Alaska Workshop, *Science,* **204,** 1413

Jones, W. L., Boggs, D. H., Bracalente, E. M., Brown, R. A., Guymer, T. H., Shelton, D. and Schroeder, L. C., 1981, Evaluation of the SEASAT wind scatterometer, *Nature,* **294,** 704

Justice, C. O., Townshend, J. R. G., Holben, B. N. and Tucker, C. J., 1985, Analysis of the phenology of global vegetation using meteorological satellite data. *International Journal of Remote Sensing,* **6,** 1271

Klemas, V., Srna, R., Treasure, W. M. and Rogers, R., 1974, Satellite studies of turbidity and circulation patterns in Delaware Bay. *Proceedings of the American Society of Photogrammetry Fall Convention and Symposium on Remote Sensing in Oceanography,* Lake Buena Vista, October 1973 (Falls Church, VA: American Society of Photogrammetry and Remote Sensing) p. 848

Kung, R. T. V. and Itzkan, I., 1976, Absolute oil fluorescence conversion efficiency. *Applied Optics,* **15,** 409

Labs, D. and Neckel, H., 1967, The absolute radiation intensity of the centre of the Sun disc in the spectral range 3288–12480 Å. *Zeitschrift für Astrophysik,* **65,** 133

Labs, D. and Neckel, H., 1968, The radiation of the solar photosphere. *Zeitschrift für Astrophysik,* **69,** 1

Labs, D. and Neckel, H., 1970, Transformation of the absolute solar radiation data into the International Practical Temperature Scale of 1968. *Solar Physics,* **15,** 79

Lauritson, L., Nelson, G. J. and Porto, F. W., 1979, Data extraction and calibration of TIROS-N/NOAA radiometers. NOAA Technical Memorandum NESS 107, US Department of Commerce, Washington, DC

Lillesand, T. M. and Kiefer, R. W., 1987, *Remote Sensing and Image Interpretation* (New York: Wiley)

Lodge, D. W. S., 1981, The SEASAT-1 synthetic aperture radar: introduction, data reception and processing. In *Remote Sensing in Meteorology, Oceanography and Hydrology,* edited by A. P. Cracknell (Chichester: Ellis Horwood) p. 335

Longuet-Higgins, M. S., 1952, On the statistical distribution of the heights of sea waves, *Journal of Marine Research,* **11,** 245

McCord, H. L., 1962, The equivalence among three approaches to deriving synthetic array patterns and analysing processing techniques, *IRE Transactions on Military Electronics,* **MIL-6,** 116

MacFarlane, N. and Robinson, I. S., 1984, Atmospheric correction of Landsat MSS data for a multidate suspended sediment algorithm. *International Journal of Remote Sensing,* **5,** 561

McKenzie, R. L. and Nisbet, R. M., 1982, Applicability of satellite-derived sea-surface temperatures in the Fiji region. *Remote Sensing of the Environment,* **12,** 349

MacPhee, S. B., Dow, A. J., Anderson, N. M. and Reid, D. B., 1981, Aerial hydrography laser bathymetry and air photo interpretation techniques for obtaining inshore hydrography. *XVIth International Congress of Surveyors,* Montreux, August 1981, Paper 405.3

Maracci, G. C., 1978a, Measurement of atmospheric parameters: AGRESTE Project – Agricultural resources investigations in Northern Italy and Southern France. Commission of the European Communities, Ispra, **159**

Maracci, G. C., 1978b, Use of captive balloons for spectral signature measurements: AGRESTE Project — Agricultural resources investigations in Northern Italy and Southern France. Commission of the European Communities, Ispra, **159**

Maresca, J. W. and Barnum, J. R., 1977, Measurement of oceanic wind speed from HF scatter by skywave radar. *IEEE Transactions on Antennas and Propagation,* **AP-25**, 132

Maul, G. A., 1981, Application of GOES visible-infrared data to quantifying mesoscale ocean surface temperatures. *Journal of Geophysical Research,* **86**, 8007

Muirhead, K. and Cracknell, A. P., 1986, Review article: Airborne lidar bathymetry. *International Journal of Remote Sensing,* **7**, 597

Munday, J. C. and Alfoldi, T. T., 1979, LANDSAT test for diffuse reflectance models for aquatic suspended solids measurement. *Remote Sensing of the Environment,* **8**, 169

NASA, 1976, *Landsat Data User's Handbook.* Document No. 76SDS4258 (Greenbelt, MD: National Aeronautics and Space Administration)

Needham, B. H., 1983, NOAA's activities in the field of marine remote sensing. In *Remote Sensing Applications in Marine Science and Technology,* edited by A. P. Cracknell (Dordrecht: D. Reidel) p. 17

Offiler, D., 1983, Surface wind vector measurements from satellites. In *Remote Sensing Applications in Marine Science and Technology,* edited by A. P. Cracknell (Dordrecht: D. Reidel) p. 169

O'Neil, R. A., Buga-Bijunas, L. and Rayner, D. M., 1980, Field performance of a laser fluorosensor for the detection of oil spills. *Applied Optics,* **19**, 863

O'Neil, R. A., Hoge, F. E. and Bristow, M. P. F., 1981, The current status of airborne laser fluorosensing. *Proceedings of 15th International Symposium on Remote Sensing of Environment,* Ann Arbor, MI, May 1981, p. 379

Østrem, G., 1981, The use of remote sensing in hydrology in Norway. In *Remote Sensing in Meteorology, Oceanography and Hydrology,* edited by A. P. Cracknell (Chichester: Ellis Horwood) p. 258

Phulpin, T. and Deschamps, P.Y., 1980, Estimation of sea surface temperature from AVHRR infrared channels measurements. In *Coastal and Marine Applications of Remote Sensing,* edited by A. P. Cracknell (Reading: Remote Sensing Society) p. 47

Prabhakara, G., Dalu, G. and Kunde, V. G., 1974, Estimation of sea surface temperature from remote sensing in the 11- to 13-μm window region. *Journal of Geophysical Research,* **79**, 5039

Rao, P. K., Smith, W. L. and Koffler, R., 1972, Global sea surface temperature distribution determined from an environmental satellite. *Monthly Weather Review,* **100**, 10

Rayner, D. M., 1979, A laboratory study of the potential of time resolved laser fluorosensors. Report No. BY-79-2(RC), Division of Biological Sciences, National Research Council, Ottawa

Rhind, D. W. and Mounsey, H., 1991, *Understanding Geographic Information Systems* (London: Taylor & Francis)

Rice, S. O., 1951, Reflection of electromagnetic waves from slightly rough surfaces. In *Theory of Electromagnetic Waves,* edited by M. Kline (New York: Interscience)

Ross, D. and Jones, W. L., 1978, On the relationship of radar backscatter to windspeed and fetch. *Boundary-Layer Meteorology,* **13**, 151

Sathyendranath, S. and Morel, A., 1983, Light emerging from the sea – interpretation and uses in remote sensing. In *Remote Sensing Applications in Marine Science and Technology,* edited by A. P. Cracknell (Dordrecht: D. Reidel) p. 323

Saunders, R. W., 1982, Methods for the detection of cloudy pixels. *Remote Sensing and the Atmosphere: Proceedings of the Annual Technical Conference of the Remote Sensing Society,* Liverpool, December 1982 (Reading: Remote Sensing Society) p. 256

Schneider, S. R., McGinnis, D. F. and Gatlin, J. A., 1981, Use of NOAA/AVHRR visible and near-infrared data for land remote sensing. NOAA Technical Report NESS 84, US Department of Commerce, Washington, DC

Schroeder, L. C., Boggs, D. H., Dome, G., Halberstam, I. M., Jones, W. L., Pierson, W. J. and Wentz, F. J., 1982, The relationship between wind vector and normalised radar cross section used to derive SEASAT-A satellite scatterometer winds. *Journal of Geophysical Research,* **87C**, 3318

Shearman, E. D. R., 1981, Remote sensing of ocean waves, currents and surface winds by dekametric radar. In *Remote Sensing in Meteorology, Oceanography and Hydrology,* edited by A. P. Cracknell (Chichester: Ellis Horwood) p. 312

Shearman, E. D. R., Sandham, W. A., Bramley, E. N. and Bradley, P. A., 1979, Ground-wave and sky-wave sea-state sensing experiments in the UK. Agard Conference Special Topics in HF Propagation, Lisbon, Agard Conference Proceedings No. 263

Sheffield, C., 1981, *Earthwatch: a survey of the Earth from space* (London: Sidgwick and Jackson)

Sheffield, C., 1983, *Man on Earth* (London: Sidgwick and Jackson)

Sidran, M., 1980, Infrared sensing of sea surface temperature from space. *Remote Sensing of the Environment,* **10**, 101

Singh, S. M. and Cracknell, A. P., 1979, Analysis of LANDSAT scene of 27th June 1977 for EURASEP test site and correlation with OCS results. *Proceedings of OCS Workshop,* October 1979, edited by B. M. Sørensen (Ispra: Joint Research Centre) p. 171

Singh, S. M., Cracknell, A. P. and Charlton, J. A., 1983, Comparison between CZCS data from 10 July 1979 and simultaneous *in situ* measurements for south-eastern Scottish waters. *International Journal of Remote Sensing,* **4**, 755

Singh, S. M., Cracknell, A. P. and Fiúza, A. F. G., 1985, The estimation of atmospheric corrections to one-channel (11 μm) data from the AVHRR; simulation using AVHRR/2. *International Journal of Remote Sensing,* **6**, 927

Singh, S. M., Cracknell, A. P. and Spitzer, D., 1985, Evaluation of sensitivity decay of Coastal Zone Colour Scanner (CZCS) detectors by comparison with *in situ* near-surface radiance measurements. *International Journal of Remote Sensing,* **6**, 749

Singh, S. M. and Warren, D. E., 1983, Sea surface temperatures from infrared measurements. In *Remote Sensing Applications in Marine Science and Technology,* edited by A. P. Cracknell (Dordrecht: D. Reidel) p. 231

Smart, P. L. and Laidlaw, I. M. S., 1977, An evaluation of some fluorescent dyes for water tracing. *Water Resources Research,* **13**, 15

Smith, R. C. and Wilson, W. H., 1980, Ship and satellite bio-optical research in the California Bight. *COSPAR/SCOR/IUCRM, Symposium on oceanography from space,* Venice, May 1980

Sturm, B., 1981, The atmospheric correction of remotely sensed data and the quantitative determination of suspended matter in marine water surface layers. In *Remote Sensing in Meteorology, Oceanography and Hydrology,* edited by A. P. Cracknell (Chichester: Ellis Horwood) p. 163

Sturm, B., 1983, Selected topics of Coastal Zone Color Scanner (CZCS) data evaluation. In *Remote Sensing Applications in Marine Science and Technology,* edited by A. P. Cracknell (Dordrecht: D. Reidel) p. 137

Thekaekara, M. P., Kruger, R. and Duncan, C. H., 1969, Solar irradiance measurements from a research aircraft. *Applied Optics,* **8,** 1713

Thomas, D. P., 1981, Microwave radiometry and applications. In *Remote Sensing in Meteorology, Oceanography and Hydrology,* edited by A. P. Cracknell (Chichester: Ellis Horwood) p. 357

Tournier, B., 1978, Determination des temperatures de surface de la mer à partir des mésures radiometriques satellitaires. Utilisation pour l'Océanologie des Satellites d'Observation de la Terre, Journées Nationales d'Etudes, Brest, Fevrier 1978, CNEXO, Paris, p. 33

Townsend, W. F., 1980, An initial assessment of the performance achieved by the SEASAT-1 radar altimeter. *IEEE Journal of Oceanographical Engineering,* **OE-5,** 80

Valerio, C., 1981, Airborne remote sensing experiments with a fluorescent tracer. In *Remote Sensing in Meteorology, Oceanography and Hydrology,* edited by A. P. Cracknell (Chichester: Ellis Horwood) p. 218

Valerio, C., 1983, Airborne remote sensing and experiments with fluorescent tracers. In *Remote Sensing Applications in Marine Science and Technology,* edited by A. P. Cracknell (Dordrecht: D. Reidel) p. 383

Ward, J. F., 1969, Power spectra from ocean movements measured remotely by ionospheric radio backscatter. *Nature,* **223,** 1325

Weinreb, M. P. and Hill, M. L., 1980, Calculation of atmospheric radiances and brightness temperatures in infrared window channels of satellite radiometers. NOAA Technical Report NESS 80, US Department of Commerce, Rockville, MD

Werbowetzki, A., 1981, Atmospheric sounding user's guide. NOAA Technical Report NESS 83, US Department of Commerce, Washington, DC

Wilson, H. R., 1981, Elementary ideas of optical image processing. In *Remote Sensing in Meteorology, Oceanography and Hydrology,* edited by A. P. Cracknell (Chichester: Ellis Horwood), p. 114

Wilson, S. B. and Anderson, J. M., 1984, A thermal plume in the Tay estuary detected by aerial thermography. *International Journal of Remote Sensing,* **5,** 247

Wurtele, M. G., Woiceshyn, P. M., Peteherych, S., Borowski, M. and Appleby, W. S., 1982, Wind direction alias removal studies of SEASAT scatterometer-derived wind fields. *Journal of Geophysical Research,* **87C,** 3365

Wyatt, L., 1983, The measurement of oceanographic parameters using dekametric radar. In *Remote Sensing Applications in Marine Science and Technology,* edited by A. P. Cracknell (Dordrecht: D. Reidel) p. 183

Zwick, H. H., Neville, R. A. and O'Neil, R. A., 1981, A recommended sensor package for the detection and tracking of oil spills. *Proceedings of an EARSeL-ESA Symposium,* Voss, Norway, May 1981. **ESA SP-167,** p. 77

Appendix I Bibliography

The following are not cited specifically in the text but are general references which readers may find useful as sources of further information or discussion.

Allan, T. D., 1983, *Satellite microwave remote sensing* (Chichester, UK: Ellis Horwood)

Barrett, E. C. and Curtis, L. F., 1982, *Introduction to environmental remote sensing* (New York: Chapman and Hall)

Browning, K. A., 1979, The FRONTIERS plan: a strategy for using radar and satellite imagery for very-short-range precipitation forecasting. *Meteorological Magazine,* **108**, 161

Carter, D. J., 1986, *The remote sensing sourcebook: a guide to remote sensing products, services, facilities, publications and other materials* (London: Kogan Page, McCarta)

Cracknell, A. P., 1981, *Remote Sensing in Meteorology, Oceanography and Hydrology* (Chichester, UK: Ellis Horwood)

Cracknell, A. P., 1983, *Remote Sensing Applications in Marine Science and Technology* (Dordrecht: D. Reidel)

Cracknell, A. P., Hayes, L. W. B. and Huang, W. G. (Eds), 1990, *Remote Sensing Yearbook* (London: Taylor & Francis)

Curran, P. J., 1985, *Principles of remote sensing* (New York: Longman)

Drury, S. A., 1987, *Image interpretation in geology* (London: George Allen & Unwin)

Gonzalez, R. C. and Wintz, P., 1977, *Digital image processing* (Reading, MA: Addison-Wesley)

Griersmith, D. C. and Kingwell, J., 1988, *Planet under scrutiny – an Australian remote sensing glossary* (Canberra: Australian Government Publishing Service)

Hall, D. K. and Martinec, J., 1985, *Remote sensing of ice and snow* (London: Chapman and Hall)

Houghton, J. T., 1977, *The physics of atmospheres* (Cambridge: Cambridge University Press)

Houghton, J. T., Taylor, F. W. and Rodgers, C. D., 1984, *Remote sounding of atmospheres* (Cambridge: Cambridge University Press)

Hyatt, E., 1988, *Keyguide to information sources in remote sensing* (London: Mansell)

Kennie, T. J. M. and Matthews, M. C., 1985, *Remote sensing in civil engineering* (Glasgow and London: Surrey University Press)

Lo, C. P., 1986, *Applied remote sensing* (Harlow, UK: Longman)

Mason, B. D., 1981, METEOSAT — Europe's contribution to the global weather observing system. In *Remote Sensing in Meteorology, Oceanography and Hydrology,* edited by A. P. Cracknell (Chichester, UK: Ellis Horwood), p. 56

Maul, G. A., 1985, *Introduction to satellite oceanography* (Dordrecht: Martinus Nijhoff)

Merrick, G., 1980, The uses of satellites. I. The space shuttle — today's system for tomorrow's world. *Royal Society of Arts Journal,* **128**, 797

Miller, D. E., 1980, The uses of satellites. II. Weather satellites. *Royal Society of Arts Journal,* **128**, 813

Muller, J-P. (Ed.), 1988, *Digital Image Processing in Remote Sensing* (London: Taylor & Francis)

Murtha, P. A. and Harding, R. A., 1984, *Renewable resources management: applications of remote sensing* (Falls Church, VA: American Society of Photogrammetry and Remote Sensing)

Rabchevsky, G. A., 1984, *Multilingual dictionary of remote sensing and photogrammetry* (Falls Church, VA: American Society of Photogrammetry and Remote Sensing)

Reynolds, M., 1977, Meteosat's imaging payload. *ESA Bulletin,* **11**, 28

Robinson, I. S., 1985, *Satellite oceanography* (Chichester, UK: Ellis Horwood)

Sabins, F. F., 1986, *Remote sensing: principles and interpretation* (New York: John Wiley)

Schanda, E., 1976, *Remote sensing for environmental sciences* (Heidelberg: Springer-Verlag)

Schanda, E., 1986, *Physical fundamentals of remote sensing* (Heidelberg: Springer-Verlag)

Shirvanian, D., 1988, *European Space Directory* (Paris: Sevig)

Siegal, B. S. and Gillespie, A. R., 1980, *Remote sensing in geology* (New York: John Wiley)

Slater, P. N., 1980, *Remote sensing optics and optical systems* (Reading, MA: Addison-Wesley)

Stewart, R. H., 1985, *Methods of satellite oceanography* (Berkeley, CA: University of California Press)

Swain, P. H. and Davis, S. M., 1978, *Remote sensing: the quantitative approach* (New York: McGraw-Hill)

Szekielda, K. H., 1986, *Satellite remote sensing for resources development* (London: Graham & Trotman)

Taillefer, Y., 1982, *Receuil de terminologie spatiale/A glossary of space terms* (Paris: European Space Agency)

Townshend, J. R. G., 1981, *Terrain analysis and remote sensing* (London: George Allen & Unwin)

Trevett, J. W., 1986, *Imaging radar for resources surveys* (London: Chapman and Hall)

Ulaby, F. T., Moore, R. K. and Fung, A. F., 1981, *Microwave remote sensing: active and passive: Volume 1, MRS fundamentals and radiometry* (Reading, MA: Addison-Wesley)

Ulaby, F. T., Moore, R. K. and Fung, A. F., 1982, *Radar remote sensing and surface scattering and emission theory* (Reading, MA: Addison-Wesley)

Ulaby, F. T., Moore, R. K. and Fung, A. F., 1986, *From theory to applications* (London: Adtech)

Vertsappen, H. Th., 1977, *Remote sensing in geomorphology* (Amsterdam: Elsevier)

Widger, W. K., 1966, *Meteorological satellites* (New York: Holt, Rinehart & Winston)

Yates, H. W. and Bandeen, W. R., 1975, Meteorological applications of remote sensing from satellites. *Proceedings IEEE,* **63**, 148

Appendix II Sources of remotely sensed data

The following information is extracted from *Remote Sensing Yearbook 1990*, edited by A. P. Cracknell *et al.* (London: Taylor & Francis)

International LANDSAT data distribution centres

Argentina
Comision Nacional de Investigaciones
  Espaciales
Centro de Procesamiento
Avenue Dorrego 4010
1425 Buenos Aires
Tel. Buenos Aires 7225108
Tlx. 17511 Lanba

Australia
Australia Landsat Station
22–36 Oatley Court
PO Box 28
Belconnen
ACT 2616
Tel. 062/515411
Tlx. 61510

Brazil
INPE-DGI
Caixa Postal 01
Cachoeira Paulista-Cep 12,630
Sao Paulo
Tel. Sao Paulo 611507
Tlx. 122160 INPE

Canada
Canada Centre for Remote Sensing
User Assistance and Marketing Unit
717 Belfast Road
Ottawa

Ontario K1A OY7
Tel. Ottawa 9951210
Tlx. 0533777

China
Academia Sinica
Landsat Ground Station
Beijing
Tel. Beijing 284861
Tlx. 210222 ASCHI

European Space Agency
ESA-ESRIN
Earthnet User Services
Via Galileo Galilei
00044 Frascati
Italy
Tel. Rome 9401360 or Rome 94011
Tlx. 610637 ESRIN
(Data collected at Fucino, Kiruna, and
  Masapalomas are ordered through
  ESA-ESRIN.)

India
National Remote Sensing Agency
Department of Space
Balanager
500037 Hyderabad
Andhra Pradesh
Tel. 262572–77
Tlx. 0155–522, 0155–6522

Indonesia
Indonesian National Institute of
  Aeronautics and Space (Lapan)
JL Permuda Persil No. 1
PO Box 3048
Jakarta
Tlx. 49175 LAPAN

Japan
Remote Sensing Technology Center of
  Japan (Restec)
Uni-Roppongi Bldg
7–15–17 Roppongi
Minato-ku
Tokyo 106
Tel. Tokyo 4031761
Tlx. 022426780 RESTEC

Pakistan
Pakistan Space and Upper Atmosphere
  Research Commission (Suparco)
43–1/P-6 Pecks
PO Box 3125
Karachi-Az
Tlx. 25720 SPACE

Saudi Arabia
King Abdulaziz City for Science and
  Technology
PO Box 6086
Riyadh 11442
Tlx. 201590 Sancst Space

South Africa
National Institute for
  Telecommunications Research
Attn: Satellite Remote Sensing Center
PO Box 3718
Johannesburg 2000
Tel. 012/265271, 011/642 4693
Tlx. 3–21005

Thailand
Remote Sensing Division
National Research Council of Thailand
196 Phahonyothin Road
Bangkhen
Bangkok 10900
Tel. Bangkok 579 0116
Tlx. 82213 NARECOU
Telegram. NRC Bangkok

UK
Environmental Remote Sensing Unit
British Aerospace (Space Systems)
FPC 311, Box 5
Filton, Bristol BS12 7QW
Tel. 0272 366832/366416
Tlx. 449452

USA
Earth Observation Satellite Company
  (EOSAT)
c/o Eros Data Center
Sioux Falls
SD 57198
Tel. 605/5942291, 800/3672801
Tlx. 910–668–0310 EDCSFL

Reception and distribution of SPOT data

Australia
NATMAP
Division of National Mapping
PO Box 31
Belconnen
ACT 2616

Brazil
Institute de Pesquisas Espaciais (INPE)
C.P. S15

Sao José dos Campos
12200 Sao Paulo

Canada
Canada Center for Remote Sensing
2464 Sheffield Road
Ottawa
Ottawa K1A OY7

China
Space Science and Technology Center
Chinese Academy of Science
Beijing

India
National Remote Sensing Agency
Department of Space
Balanager
Hyderabad
500037

Japan
NASDA

2-4-1, Hamamatsu-Cho
Minato-Ku
Tokyo 105

Pakistan
SUPARCO
PO Box 3125
Karachi

Saudi Arabia
King Abdulaziz City for Science and
 Technology
PO Box 6086
Riyadh 11442

SPOT data distributors

Argentina
Centro Nacional de Investigaciones
 Espaciales
Centro de Teleobservación
Av. del Libertador 1513
Vincente Lopez 1638
Buenos Aires

Austria
Beckel Satellitenbilddaten
Marie-Louisen Strasse
4820 Bad Ischl

Belgium
Services de la Programmation de la
 Politique Scientifique
8 rue de la Science
1040 Bruxelles

Bolivia
Centro de Investigación y Aplicación de
 Sensores Remotos
Casilla de correo 2729
La Paz

Brazil
Sensora
Avenida Sernambetiba
NR 4446
Rio de Janeiro
CEP 22600

Canada
DIGIM
1100 Blvd Dorchester West
Montreal
Quebec H3B 4P3

Chile
Servicio Aerofotogramétrico de la
 Fuerza Aerea
Casilla 67 correo los cerillos
Santiago

Denmark
Plancenter Fyn A/S
Overgade 32
5000 Odense C

Egypt
Remote Sensing Center
101 Kasr EL Eini Street
Cairo

Finland
National Board of Survey
Pasilan Virastukeskus
Opastinsilta 12
00521 Helsinki 52

France
SPOT Image
18 Avenue Edouard-Belin
31055 Toulouse Cédex
France

Germany
Deutsche Forchungs- und
    Versuchanstalt für Luft- und
    Raumfahrt
Oberpfaffenhofen
8031 Wessling

Hungary
Földmeresi Intezet
Guszev v. 19
1051 Budapest

Ireland
Remote Sensing Geological and
    Environmental Services
Environmental Resources Analysis Ltd
187 Pearse Street
Dublin 2

Israel
Interdisciplinary Center for
    Technological Analysis and
    Forecasting
Ramat-Aviv
Tel-Aviv 69978

Italy
Telespazio
Département Commercial
Via Alberto Bergamini 50
00159 Rome

Japan
Remote Sensing Technology Center
Uni Roppongi Bldg
7–15–17
Roppongi
Minato-ku
Tokyo 106

Malawi
Geoservices Ltd
PO Box 30305
Lilongwe 3

Malaysia
Terra Control Technologies
Sdn. Bhd
Godown 3, Banguman Nupro
Jalan Brickfield
50470 Kuala Lumpur

Mexico
Instituto Nacional de Estadistica
    Geografia e Informatica
San Antonio Abad 124
Mexico 8 DF

Nepal
National Remote Sensing Center
PO Box 3103
Kathmandu

Netherlands
National Lucht en
    Ruimtevaarlaboratorium
PO Box 90502
BM Amsterdam 1006

Nigeria
Danz Surveys and Consultants
24 Oyekan Road
Lagos

Norway
Fjellanger Wideroe A/S
PO Box 2916
7001 Trondheim

Peru
Oficina Nacional de Evaluación de
    Recursos Naturales
355 calle 17 Urb El Palomar
San Isidro
Lima

Philippines
Natural Resources Management Center
PO Box AC
Quezon City 493

Poland
Geokart
2/4 rue Jasna
00950 Warsaw

Portugal
Geometral
Ave Cons. Barjona De Freitas 20-A
1500 Lisbon

South Africa
Council for Scientific and Industrial
  Research
Foundation for Research Development
PO Box 395
0001 Pretoria

Spain
Instituto Geografico Nacional
General Ibanez de Ibero 3
Madrid 3

Sweden
Satimage
PO Box 816
98128 Kiruna

Switzerland
Bundesamt für Landestopographie
Seftigenstr. 264
3084 Wabern

Taiwan
Center for Space and Remote Sensing
  Research
National Central University
320 Chung-Li

Thailand
National Research Council
196, Phahonyothin Road
10900 Bangkok

United Kingdom
Nigel Press Associates Ltd
Edenbridge
Kent
TN8 6HS

United Kingdom
National Remote Sensing Centre
Space Department
Royal Aerospace Establishment
Farnborough
Hants GU14 6TD

USA
SPOT Image Corp.
1897 Preston White Drive
Reston
VA 22091

Venezuela
Fundación Instituto de Ingeneiría
  (C.P.D.I.)
Edo Mirando Apartado 40200
1040 Caracas A

Yugoslavia
Rudarski Institute Beograd
Remote Sensing Department
Batajnicki put 2
11081 Zemun

Points of contact for data from other satellites

US Environmental Satellites
Satellite Data Services Division
NOAA/EDIS/NCC
World Weather Building, Room 100
Washington, DC 20233
USA
Tel. 3017638111

NASA Experimental Satellites
National Space Science Data Center
NASA/Goddard Space Flight Center
Code 601
Greenbelt, MD 20771
USA
Tel. 3013446695

US Defense Meteorological Satellite
Space Science & Engineering Center
DMSP Archive
University of Wisconsin
1225 West Dayton Street
Madison, WI 53706
USA
Tel. 6082625335

The prime international data distribution center for manned satellite sensor data (other than Space Shuttle)
US Department of Commerce
NOAA
National Earth Satellite Service
EROS Data Center
Sioux Falls
SD 57198
USA

Spacelab data
European Space Agency
8–10 Rue Mario Nikis
75738 Paris 15
France

European (Meteosat) Satellite Data
Meteorological Data Management
    Department
European Space Operations Center

Robert-Boschstrasse 5
6100 Darmstadt
Germany
Tel. 061518861

Japanese Satellites
Data Processing Department
Meteorological Satellite Center
3–235 Nakayoto Kiyose
Tokyo 180
Tel. 0304249311

Indian Meteorological Satellite INSAT I-B
Director General
Indian Meteorological Department
Mausam Bhauan
Lodi Road
New Delhi

GMS Data
Mr M. Koga
Japan Weather Association
Kaiji Center Bldg.
No. 4–5, Koojimachi
Chiyoda-Ku
Tokyo 102
Japan
Tel. 3/230 0381
Tlx. 232 4863 JWAWNJ

National points of contact for Earthnet

Austria
Dr E. Mondre
Austrian Space & Solar Agency
Garnisongasse 7
1090 Vienna
Tel. Vienna 4381770
Tlx. 1165060 ASSA

Belgium
J. Theatre
Institute Géographique National
13 Abbaye de la Cambre

1051 Brussels
Tel. Brussels 6486480
Tlx. 24367 IGNGI

Denmark
Prof P. Gudmansen
Electromagnetic Institute
DTH Building 348
2800 Lyngby
Tel. Lyngby 881444
Tlx. 37529 DTHDIA

Finland
Mr J. Paavilainen
National Board of Survey
Pasila Office Centre
Postilokero 84
00521 Helsinki 52
Tlx. 125254 MAP

France
Mr G. Bonne
GDTA Centre Spatiale de Toulouse
18 Avenue Edouard Belin
31055 Toulouse
Tel. Toulouse 274281
Tlx. 531081 CNEST B

Germany
H. Engel
DFVLR
Hauptabteilung Raumflugbetrieb
8031 Oberpfaffenhofen
Tel. 08153/28740
Tlx. 52742–03 SOC

Hungary
Dr E. Csato
Institute of Geodesy, Cartography
  and Remote Sensing
Guszev u.19
1051 Budapest
Tlx. 224964

Ireland
Dr B. O'Donnell
National Board for Science &
  Technology
Shelbourne House
Shelbourne Road
Dublin 4
Tel. 683311
Tlx. 30327 NBST

Italy
Mr Maranese
Telespazio
Via Bergamini 50
00158 Rome
Tel. 49872523
Tlx. 610654 TSPZRO

Netherlands
Dr F. Van der Laan
NLR-NPOC
National Aerospace Laboratory
PO Box 153
8300 AD Emmeloord
Tlx. 11118 NLR AA

Norway
Mr Rolf-Terje Enoksen
Tromsø Telemetry Station
PO Box 387
9001 Tromsø
Tel. Tromsø 84817
Tlx. 64025 SPACE

Poland
Prof B. Ney
Glowny Urzad Geodezji i
  Kartografii
ul. Jasna 2/4 Skr. Pt. 145
00950 Warsaw
Tlx. 812770 GEOK

Romania
Dr Constantin Teodorescu
The Romanian Commission for Space
  Activities
15 Constantin Mille St
Bucharest
Tlx. 11575 ICST

Spain
Mr R. Barco
INTA
Pintor Rosales 34
Madrid 8
Tel. Madrid 2479800
Tlx. 23495 INVES

Sweden
Mrs I. Kuukasjarvi
SATIMAGE
PO Box 816
S-981 28 Kiruna
Tel. Kiruna 12140
Tlx. 8761 SATIMA

Switzerland
Mr Ch. Eidenbenz
Bundesamt für Landestopographie
Seftigenstrasse 264
3084 Wabern
Tel. Berne 549111
Tlx. 33385 LATOP

United Kingdom
Mr M. Hammond
National Remote Sensing Center
Royal Aircraft Establishment
Farnborough
Hants. GU4 6TD
Tel. Farnborough 24461
Tlx. 858442 PE MOD

AVHRR data distributors in Europe

Denmark
Mr T. Rye Nielsen
Meteorological Institute
Observatory for Space Research
Rudeskov
3460 Birkeroed

France
Dr Pascal Brunel
Centre de Météorologie Spatiale
BP 147
22302 Lannion Cédex

Germany
Dipl.-Met. M. Eckhardt
Freie Universität Berlin
Institut für Meteorologie
Dietrich-Schafer-Weg 6–10
1000 Berlin 41

Germany
Mr K. Reigniger
DFVLR

Oberpfaffenhofen
8031 Wessling

Norway
Mr Rolf-Terje Enoksen
Tromsø Telemetry Station
PO Box 387
9001 Tromsø

United Kingdom
Mr M. Boswell
RAE Lasham
Lasham Airfield
Nr Alton
Hampshire

United Kingdom
Mr P. E. Baylis
Dept of Electrical Engineering and
   Electronics
Dundee University
Dundee DD1 4HN

Countries without a Local Distributor should enquire and order via:

Ms Carla Fonti
Earthnet Programme Office
ESRIN
Via Galileo Galilei
00044 Frascati
Italy
Tel. Rome 940 1218
Tlx. 616468 EURIME

Australian sources of remotely sensed data

Australian LANDSAT Station
PO Box 28
Belconnen
ACT 2616
Tel. 062/515411
Tlx. 61510

Australian LANDSAT Station
CSIRO Compound
Health Road
Alice Springs
NT 5750
Tel. 089/523353
Tlx. 81386

New South Wales Lands Department
Map Sales
22–23 Bridge Street
Sydney
NSW 2000
Tel. 02/20579
Cable. LANDEP SYDNEY

Sunmap Aerial Photography Section
Department of Mapping and Surveying
11th Floor, Watkins Place
288 Edwards Street
Brisbane
Qsld 4000
Tel. 07/277784
Tlx. 41412

Central Map Agency
Department of Lands and Surveys
Cathedral Avenue
Perth
WA 6000
Tel. 09/3231349
Tlx. 93784

Mapland
Department of Lands
12 Pirie Street
Adelaide
SA 5000
Tel. 08/2272675
Tlx. 82827

Map Sales
Department of Conservation, Forest
   and Lands
25 Spring Street
Melbourne
Vic. 3000
Tel. 03/6513024
Tlx. 32636

Tasmanian Government Publications
   Centre
134 Macquarie Street
Hobart
Tas. 7000
Tel. 02/303382
Tlx. 57250

Survey Mapping Division
Department of Lands and Housing
Moonta House
Mitchell Street
Darwin
NT 5790
Tel. 089/897572
Tlx. 85453

Technical and Field Surveys
250 Pacific Highways
Crows Nest
NSW 2065
Tel. 02/4383700
Tlx. 21822

Sources of aerial photography in the United Kingdom and the USA

Government sources in the United Kingdom

Air Photo Unit
Department of Environment
6th Floor
Prince Consort House
Albert Embankment
London SE1 7TS

Central Register of Air Photographs for Wales
Welsh Office
Cathays Park
Cardiff CF1 3NQ

The Air Photographs Officer
Central Register of Air Photography
Scottish Development Department
New St Andrews House
St James' Centre
Edinburgh EH1 3SZ

Deputy Keeper of Records
Public Record Office of Northern Ireland
66 Balmoral Avenue
Belfast 9

Ordnance Survey
Air Photo Cover Group
Romsey Road
Maybush
Southampton SO9 4DH

Clyde Surveys
Reform Road
Maidenhead
Berks SR6 8BU

Commercial sources in the United Kingdom

Meridian Air Maps Ltd
Marlborough Road
Lancing
Sussex BN15 8TT

Hunting Surveys Ltd
Elstree Way
Boreham Wood
Herts WD6 1SB

BKS Surveys Ltd
Ballycairn Road
Coleraine
Co. Londonderry BT51 3HZ

University of Cambridge
Committee for Aerial Photography
Mond Building
Free School Lane
Cambridge CB2 3RF

Government sources in the USA

Aerial Photography Field Office
ASCS-USDA
PO Box 30010
Salt Lake City
UT 84125

National Cartographic Information Center
US Geologic Survey
507 National Center
Reston
VA 22092

United States Geological Survey Aerial
   Photography
EROS Data Center
Sioux Falls
SD 57198

Commercial sources in the USA

There are a large number of relevant companies operating throughout the USA. For details refer to the academic and trade journals and local directories.

Appendix III Abbreviations and acronyms

This list includes many of the abbreviations and acronyms that one is likely to encounter in the field of remote sensing and is not limited to those used in this book. The list has been compiled from a variety of sources including:

Planet under scrutiny — an Australian remote sensing glossary, D. C. Griersmith and J. Kingwell (Canberra: Australia Government Publishing Service) 1988
Keyguide to information sources in Remote Sensing, E. Hyatt (London and New York: Mansell) 1988
Multilingual dictionary of remote sensing and photogrammetry, G. A. Rabchevsky (Falls Church, VA: American Society of Photogrammetry and Remote Sensing) 1984
Microwave remote sensing for oceanographic and marine weather-forecast models, R. A. Vaughan (Dordrecht: Kluwer) 1990.

AARS	Asian Association on Remote Sensing
ACRES	Australian Centre for Remote Sensing
ADF	Automatic Direction Finder
ADP	Automatic Data Processing
AESIS	Australian Earth Science Information System
AFC	Automatic Frequency Control
AGC	Automatic Gain Control
AIAA	American Institute of Aeronautics and Astronautics
AIT	Asian Institute of Technology
AMI	Active Microwave Instrument
AMORSA	Atmospheric and Meteorological Ocean Remote Sensing Assembly
AMSR	Advanced Microwave Scanning Radiometer
AMSU	Advanced Microwave Sounding Unit
APR	Airborne Profile Recorder; Automatic Pattern Recognition
APT	Automatic Picture Transmission
APU	Auxiliary Power Unit

ARRSTC	Asian Regional Remote Sensing Training Centre
ARSC	Australasian Remote Sensing Conference
ASPRS	American Society of Photogrammetry and Remote Sensing
ASSA	Austrian Space and Solar Agency
ATM	Airborne Thematic Mapper
ATS	Applications Technology Satellite
ATSR	Along Track Scanning Radiometer
ATSR/M	Along Track Scanning Radiometer and Microwave Sounder
AU	Astronomical Unit
AVHRR	Advanced Very High Resolution Radiometer
AVHRR/2	Advanced Very High Resolution Radiometer, 5-channel version
BARSC	British Association of Remote Sensing Companies
BCRS	Netherlands Remote Sensing Board
BNSC	British National Space Centre
BOMEX	Barbados Oceanographic and Meteorological Experiment
bpi	bits per inch
CACRS	Canadian Advisory Committee on Remote Sensing
CCD	Charged Couple Device
CCRS	Canada Centre for Remote Sensing
CCT	Computer Compatible Tape
CEOS	Committee on Earth Observation Systems
CIAF	Centro Interamericano de Fotointerpretación
CIASER	Centro de Investigación y Aplicación de Sensores Remotos
CIR	Colour Infrared Film
CLIRSEN	Centro de Levantamientos Integrados de Recursos Naturales por Sensores Remotes
CNES	Centre National D'Etudes Spatiales
CNEI	Comisión Nacional de Investigaciones Espaciales
CNR	Consiglio Nationale delle Richerche
CNRS	Centre National de la Recherche Scientifique
COPUOS	Committee on the Peaceful Uses of Outer Space
COSPAR	Committee on Space Research
CRAPE	Central Register of Aerial Photography for England
CRS	Committee of Remote Sensing (Vietnam)
CRSTI	Canadian Remote Sensing Training Institute
CRT	Cathode-Ray Tube
CRTO	Centre Régional de Télédétection de Ouagadougou
CSIRO	Commonwealth Scientific and Research Organisation (Australia)
CSRE	Centre of Studies in Resources Engineering
CW	Continuous-Wave Radar
CZCS	Coastal Zone Colour Scanner

DCP	Data Collection Platform
DCS	Data Collection System
DFVLR	Deutsche Forschungs – und Versuchsanstalt für Luft – und Raumfahrt e.v. (German Aerospace Research Establishment)
DMA	Defense Mapping Agency
DMSP	Defense Meteorological Satellite Programme (USA)
DN	Digital Number
DOD	Department of Defense (USA)
Doran	Doppler ranging
DOS	Department of Survey (USA)
DSIR	Department of Scientific and Industrial Research (New Zealand)
DST	Direct Sounding Transmission
EARSeL	European Association of Remote Sensing Laboratories
EARTHSAT	Earth Satellite Corporation
EBR	Electron Beam Recorder
ECA	Economic Commission for Africa
ECMWF	European Centre for Medium Range Weather Forecasts
EDM	Electronic Distance-Measuring Device
EEC	European Economic Community
EECF	EARTHNET ERS-1 Central Facility
ELV	Expendable Launch Vehicle
EMSS	Emulated Multispectral Scanner
EOP	Earth Observation Programme
EOPAG	ERS-1 Operation Plan Advisory Group
EOPP	Earth Observation Preparatory Programme
EOS	Earth Observing System
EOSAT	Earth Observation Satellite Company
EPO	EARTHNET Programme Office
ERB	Earth Radiation Budget
EREP	Earth Resources Experimental Package
ERIM	Environmental Research Institute of Michigan
ERISAT	Earth Science and Related Information Database
EROS	Earth Resources Observation Systems
ERS-1	Earth Resources Satellite 1 (European)
ERTS	Earth Resources Technology Satellite, later called LANDSAT
ESA	European Space Agency
ESIAC	Electronic Satellite Image Analysis Console
ESMR	Electronically Scanned Microwave Radiometer
ESOC	European Space Operations Centre
ESRIN	European Space Research Institute (Headquarters of EARTHNET Office)

ESSA	Environmental Survey Satellite
ESTEC	European Space Technology Centre
ETM	Enhanced Thematic Mapper
EUMETSAT	European Meteorological Satellite Organisation
FAO	Food and Agriculture Organization
FGGE	First GARP Global Experiment
FLI	Fluorescence Line Imager
FOV	Field of View
GAC	Global Area Coverage
GARP	Global Atmospheric Research Project
GCP	Ground Control Point
GCM	General Circulation Model
GDTA	Groupement pour Développement de la Télédétection Aérospatiale
GEMS	Global Environmental Monitoring System
GEOS/3	Geodetic Satellite
GIS	Geographic Information Systems
GMS	Geosynchronous Meteorological Satellite
GOASEX	Gulf of Alaska SEASAT Experiment
GOES	Geostationary Operational Environmental Satellite
GOFS	Global Ocean Flux Study
GOMS	Geostationary Operational Meteorological Satellite
GOS	Global Observing System
GOSSTCOMP	Global Operational Sea Surface Temperature Computation
GPS	Global Positioning System
GRD	Ground Resolved Distance
GRID	Global Resource Information Database
GSFC	Goddard Space Flight Centre
GTS	Global Telecommunication System
HBR	High Bit Rate
HCMM	Heat Capacity Mapping Mission
HCMR	Heat Capacity Mapping Radiometer
HDT	High Density Tape
HDDT	High Density Digital Tape
HIRS	High-Resolution Infrared Radiation Sounder
HIRS/2	Second generation HIRS
HRPI	High Resolution Pointable Imager
HRPT	High Resolution Picture Transmission
HRV	High Resolution Visible Scanner
ICW	Interrupted Continuous Wave
ICSU	International Council of Scientific Unions

IFDA	Institute für Angewandte Geodäsie
IFOV	Instantaneous Field-of-View
IFP	Institut Français du Pétrole
IFR	Instrument Flight Regulations
IGARSS	International Geoscience and Remote Sensing Society
IGN	Instituto Geográfico Nacional/Institut Géographique National
IGU	International Geophysical Union
IIRS	Indian Institute of Remote Sensing
IMW	Internatonal Map of the World
INPE	Instituto de Pesquisasa Espaciais
INSAT	Indian Satellite Programme
INTERCOSMOS	International Co-operation in Research and Uses of Outer Space Council
I/O	Input/Output
IOC	Intergovernmental Oceanographic Commission
IR	Infrared
IRS	Indian Remote Sensing Satellite
ISLSCP	International Satellite Land Surface Climatology Project
ISO	Infrared Space Observatory
ISPRS	International Society of Photogrammetry and Remote Sensing
ISRO	Indian Space Research Organization
ITC	International Institute for Aerospace Survey and Earth Sciences
ITOS	Improved TOS Series
JASIN	Joint Air-Sea Interaction Project
JERS-1	Japanese Earth Resources Satellite
JGOFS	Joint Global Ocean Flux Study
JPL	Jet Propulsion Laboratory (Pasadena, CA, USA)
JRC	Joint Research Centre (Ispra, Italy)
JSPRS	Japan Society of Photogrammetry and Remote Sensing
LADS	Laser Airborne Depth Sounder
LAPAN	Indonesian National Institute of Aeronautics and Space
LARS	Laboratory for Applications of Remote Sensing (Purdue University)
LBR	Laser Beam Recorder/Low Bit Rate
LFC	Large Format Camera
LFMR	Low Frequency Microwave Radiometer
LIDAR	Light Detection and Ranging
LRSA	Land Remote Sensing Assembly
LTF	Light Transfer Function
MESSR	Multispectral Electronic Self-Scanning Radiometer

MIIGAiK	Moscow Institute of Engineers for Geodesy, Aerial Surveying and Cartography
MIMR	Multi-Band Imaging Microwave Radiometer
MLA	Multispectral Linear Array
MOMS	Modular Opto-electronic Multispectral Scanner
MOP	METEOSAT Operational Programme
MOS	Marine Observation Satellite
MOS-1	Marine Observation Satellite (Japanese)
MSL	Mean Sea Level
MSR	Microwave Scanning Radiometer
MSS	Multispectral Scanner
MSU	Microwave Sounding Unit
MTF	Modulation Transfer Function
MTFF	Man Tended Free Flyer
MTI	Moving Target Indicator
NASA	National Aeronautics and Space Administration (USA)
NASDA	National Space Development Agency (Japan)
NASM	National Air and Space Museum
NEΔT	Noise Equivalent Temperature Difference
NESDIS	National Environmental Satellite Data Service (USA)
NHAP	National High Altitude Progrqam
NLR	National Lucht-en Ruimtevaartlaboratorium
NOAA	National Oceanic and Atmospheric Administration (USA)
NOAA-1	First NOAA Series Satellite, etc.
NPOC	National Point of Contact
N-ROSS	Navy Remote Ocean Sensing System (USA)
NSCAT	NASA Scatterometer
NWS	National Weather Service (USA)
OAS	Organization of American States
OBRC	On Board Range Compression
ORSA	Ocean Remote Sensing Assembly
OTV	Orbital Transfer Vehicle
PAF	Processing and Archiving Facilities
PCM	Pulse Code Modulation
PDUS	Primary Data Users Station
Pixel	Picture Element
PM	Phase Modulation
PPI	Plan Position Indicator
PRARE	Precise Range and Range Rate Equipment
PSS	Packet Switching System
RBV	Return Beam Vidicon
RECTAS	Regional Centre for Training in Aerial Surveys

RESORS	Remote Sensing Online Retrieval System (at CCRS)
RESTEC	Remote Sensing Technology Center of Japan
ROS	Radarsat Optical Scanner
RSAA	Remote Sensing Association of Australia
RSS	Remote Sensing Society
SAC	Space Applications Centre
SAF	Servicio Aerofotogramétrico de la Fuerza Aerea
SAR	Synthetic Aperture Radar
SAR-C	C-Band Synthetic Aperture Radar
SASS	SEASAT Scatterometer
SBPTC	Société Belge de Photogrammétrie, de Télédétection et de Cartographie
SCAMS	Scanning Microwave Spectrometer
SCATT-2	Scatterometer derived from ERS-1 instrument
SCS	Soil Conservation Service
SDUS	Satellite Data Users Station
SELPER	Society of Latin American Specialists in Remote Sensing
SEM	Space Environment Monitor
Shoran	Short range navigation
SIGLE	System for Information on Grey Literature in Europe
SIR	Shuttle Imaging Radar (exists as −A, −B, ...)
SLAR	Side-Looking Airborne Radar
SLR	Side Looking Radar
SMMR	Scanning Multichannel Microwave Radiometer
S/N (SNR)	Signal-to-noise ratio
SPARRSO	Bangladesh Space Research and Remote Sensing Organization
SPOT	Satellite Pour l'Observation de la Terre
SSC	Swedish Space Corporation
SSM/I	Special Sensor Microwave Imager
SST	Sea Surface Temperature
SSU	Stratospheric Sounding Unit
SUPARCO	Space and Upper Atmosphere Research Commission (Pakistan)
TDRS	Tracking Data Relay System (USA)
TIP	TIROS Information Processor
TIROS	Television and Infra-Red Observation Satellite
TIRS	Thermal Infrared Scanner
TIP	TIROS Information Processor
TM	Thematic Mapper
TOGA	Tropical Oceans Global Atmosphere
TOMS	Total Ozone Mapping Spectrometer
TOPEX	NASA Ocean Topography Experiment
TOS	TIROS Operational System

TOVS	TIROS Operational Vertical Sounder
TRF	Technical Reference File
TRSC	Thailand Remote Sensing Center
UHF	Ultra High Frequency
UN	United Nations
UNEP	United Nations Environment Programme
UNESCO	United Nations Educational, Scientific and Cultural Organisation
URSI	International Union on Radio Science
USAF	United States Air Force
USGS	United States Geological Survey
VAS	VISSR Atmospheric Sounder
VHF	Very High Frequency
VHRR	Very High Resolution Radiometer
VISSR	Visible and Infrared Spin-Scan Radiometer
VTIR	Visible and Thermal Infrared Radiometer
VTPR	Vertical Temperature Profile Radiometer
WAPI	World Aerial Photographic Index
WCDP	World Climate Data Programme
WCRP	World Climate Research Programme
WEFAX	Weather Facsimile
WISI	World Index of Space Imagery
WMO	World Meteorological Organisation
WOCE	World Ocean Circulation Experiment
WWW	World Weather Watch

Index